PASSIVE SOLAR WATER HEATERS

How to Design and Build a Batch System

PASSIVE SOLAR WATER HEATERS

How to Design and Build a Batch System

Daniel K. Reif

BRICK HOUSE PUBLISHING COMPANY
Andover, Massachusetts

Published by Brick House Publishing Co., Inc.
34 Essex Street
Andover, Massachusetts 01810

Production Credits
Editor: Jack Howell
Copy editing: Linnea Ochs
Illustrations: Barbara Putnam and Dan Reif
Cover design: Mike Fender
Book design: Kenneth J. Wilson
Typesetting: The Oxford Group

Printed in the United States of America

9, 8, 7, 6, 5, 4, 3, 2

We recommend care and adherence to standard construction safety procedures. Use welding glasses or the equivalent to protect your eyes when working with power tools. Neither the publisher nor the author takes responsibility for accidents that may occur during the building or use of any of the projects described in this book.

Library of Congress Cataloging in Publication Data

Reif, Daniel K.
Passive solar water heaters.

1. Solar heating—Passive systems. 2. Solar water heaters—Design and construction. I. Title.
TH7413.R427 1983 696'.6 82-20741
ISBN 0-931790-42-5
ISBN 0-931790-39-5 (pbk.)

ACKNOWLEDGMENTS

Throughout North America a strong and growing movement of dedicated, talented people is working toward local and national energy independence. I greatly appreciate the assistance that many of these solar pioneers have made to this book. Their hard work helps to make your job—that of the owner/builder— easier and more rewarding.

I would especially like to thank the following solar builders, designers, and researchers:

Anastas Pollock, who constructed several wall-mounted batch systems from the plans in chapter 5 and worked with me to refine the construction details and procedures;

David Bainbridge, who kept me in touch with batch heater work on the West Coast and whose fine book, *The Integral Passive Solar Water Book* (Passive Solar Institute, Davis, California, 1981), taught me much about these systems;

John Burton and Peter Zweig for use of their study on the performance of various batch designs, and for use of several of John Burton's batch heater designs;

Len Meserve, John Golder, Tom O'Brien, Marc Rosenbaum, Steve Baer, and Carl and Rael Rosato for sharing with me many of their insights into batch design and construction;

Ken Butti and John Perlin, authors of *A Golden Thread: 2500 Years of Solar Architecture and Technology*, for much of the historical material presented in chapter 2 (their book best documents the long, sporadic history of solar energy and is invaluable for all solar enthusiasts);

Jack Wiren, Terry Fenton, Marty Miller, Del Rogers, and James Saucier for their construction assistance;

Bruce Maeda and Bruce Melzer, whose computer-simulated batch performance data is used in chapter 8.

I am also grateful to:

Ed and Rosemarie Murphy and family, Bonnie Strickland, and Weepy Wooten, whose homes served as proving grounds for my prototype batch solar water heater systems (Rosemarie Murphy also provided excellent daily monitoring data on her wall-mounted batch system);

Robert Slocum and Dan Halprin for their assistance on plumbing techniques;

Charles Grandy of the Union Electric Company of St. Louis, Missouri, for use of their study on the effects of batch solar water heaters on a major utility company;

Ruth Swerdlow for her typing assistance;

And my wife, Linda, and son, Gabriel, for their patience and warm support.

CONTENTS

PASSIVE SOLAR WATER HEATERS

How to Design and Build a Batch System

BATCH WATER HEATERS

Solar works! Thousands of batch solar water heaters have been constructed from California to Maine with outstanding results. *Millions* of these collectors are in use throughout the world. Of all solar water heaters, *batch systems have proven to be the simplest to construct, easiest to maintain, and the most cost-effective.* Their simplicity makes them ideal for both do-it-yourself builders and professional contractors. Their low cost and dependability make them well suited to the needs of home owners, renters, and landlords. There are many good reasons batch solar water heaters are becoming so popular.

If you do not already have a solar water heater, you are probably paying between $200 and $600 a year to heat water for your daily household needs. *With a low-cost, easy-to-build batch solar water heater combined with basic energy conservation, you can save more than half—often two thirds—of this bill.* This saving is equal to that of many solar water heating systems costing twice as much. Year after year this adds up to a large saving, many times the original cost of the solar water heater and conservation measures.

If you live in the South or West where the climate is generally mild and there is plenty of sun, this combination of batch heater and conservation in one household of two to five people can save approximately 18.2 million Btu per year, equivalent to about 5,333 kwh of electricity, 200 gal. of oil, or 24,300 cu. ft. of natural gas. In colder, cloudier climates, performance is only slightly lower. With the collector drained to protect it from freezing during December, January, and February, savings are still approximately 14 million Btu, equivalent to roughly 4,000 kwh of electricity, 150 gal. of oil, or 18,200 cu. ft. of natural gas.* Plus, some batch heater designs can be used to provide solar house heating during the months when they are drained.

---

*Frederic Langa, "Sun on Tap: The Best We Know," Rodale's *New Shelter* magazine, May/June, 1981.

Figure 1–1.
The tank for this batch solar water heater is located in the attic, which allows the glazing to fit flush with the roof. (Designed and constructed by Nature's Way Energy Systems, Keene, New Hampshire.)

Figure 1–2.
In this housing project in Keene, New Hampshire, each of forty-one houses will be equipped with a batch water heater and passive solar space heating. (Batch designs by Nature's Way Energy Systems, Keene, New Hampshire.)

PASSIVE VERSUS ACTIVE SOLAR WATER HEATING

Solar water heaters can be divided into two categories: active and passive. Active solar water heaters rely on auxiliary power to move the collected solar heat. Drainback, draindown, and closed-loop antifreeze are examples of active solar water heating systems.

Closed-loop antifreeze systems operate by pumping a nonfreezing solution from the collectors through a heat exchanger and back to the collectors. A differential thermostat turns the pump on whenever the collector is about 15°F warmer than the water in the storage tank. When the temperature in the collector is only a few degrees warmer than the storage, the thermostat turns off the pump.

Drainback solar systems operate by circulating potable water through the collector. Freeze protection is provided by automatically draining the collector into a storage tank whenever the collector is not sufficiently warmed by the sun.

Draindown systems also circulate potable water through the collector. When the sensor detects near-freezing temperatures, the collector automatically drains into a dry well.

Each of these active solar heating systems relies on mechanical means to protect the collector from freezing. Considering Murphy's Law, there is always an element of risk.

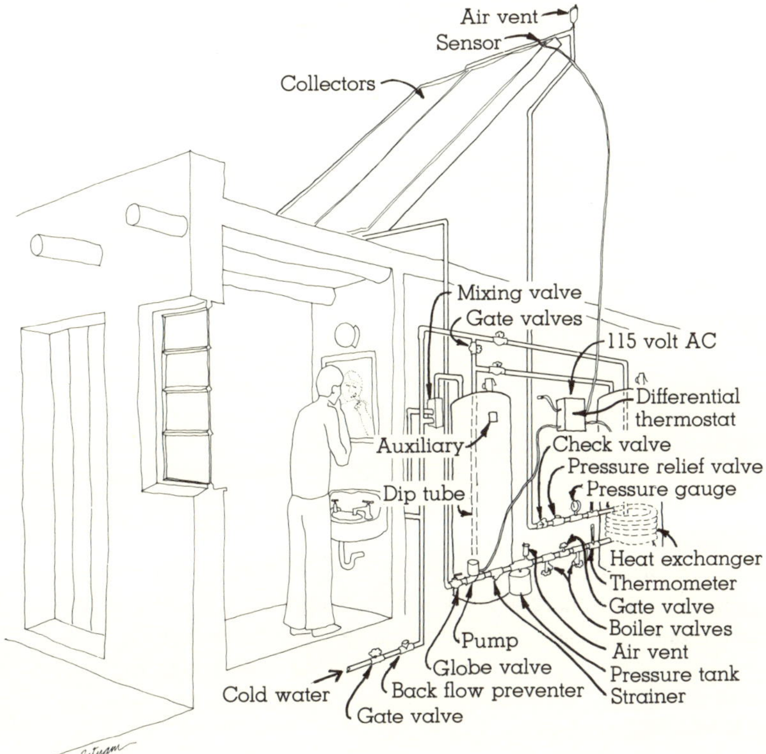

Figure 1–3.
Closed-loop antifreeze system.

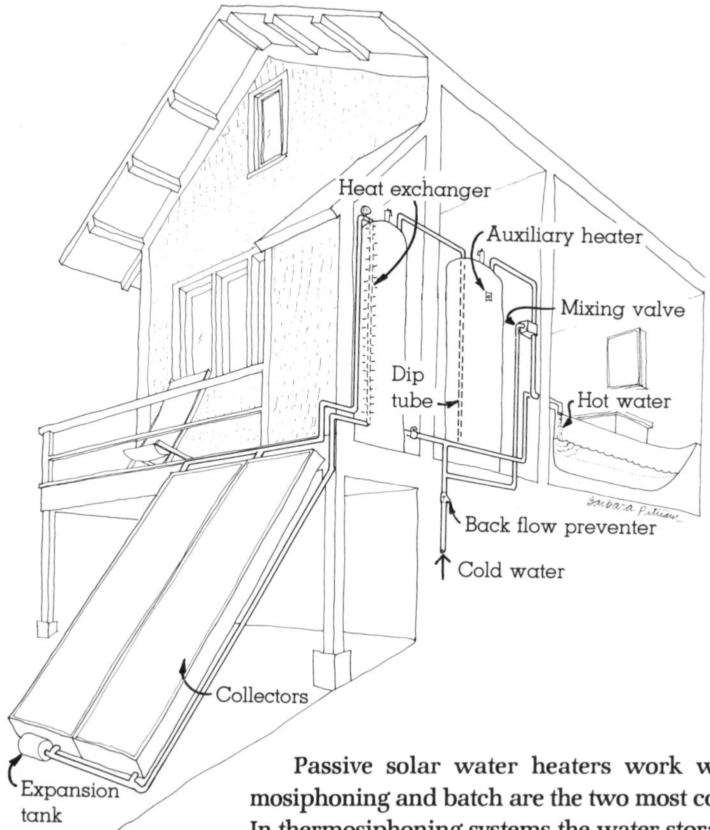

Figure 1–4.
Thermosiphoning system.

Passive solar water heaters work without any auxiliary power. Thermosiphoning and batch are the two most common types of passive water heaters. In thermosiphoning systems the water storage tank is placed above the collector. Whenever the collector is warmer than the storage tank, the warmer water rises into the storage tank and is replaced by cooler water from the bottom of the tank. Batch water heaters work by simply placing the water storage tank directly in the sun.

With passive solar water heaters, only the water moves through the pipes; there are no moving parts. It is the simplicity of passive solar water heaters, especially the batch heater, that make them low in cost, reliable, and easily constructed, and therefore the favorite of most owner/builders.

HOW BATCH HEATERS WORK

Batch solar water heaters are extremely simple. They consist of only three major components: one or more black water tanks enclosed by a combination of insulation and glazing. These collectors are called "batch" heaters because the batch of water in the tank is the heart of the system. They are also called *integral passive solar water heaters* (IPSWHs for short) because the water tank combines both solar heat collection and heat storage.

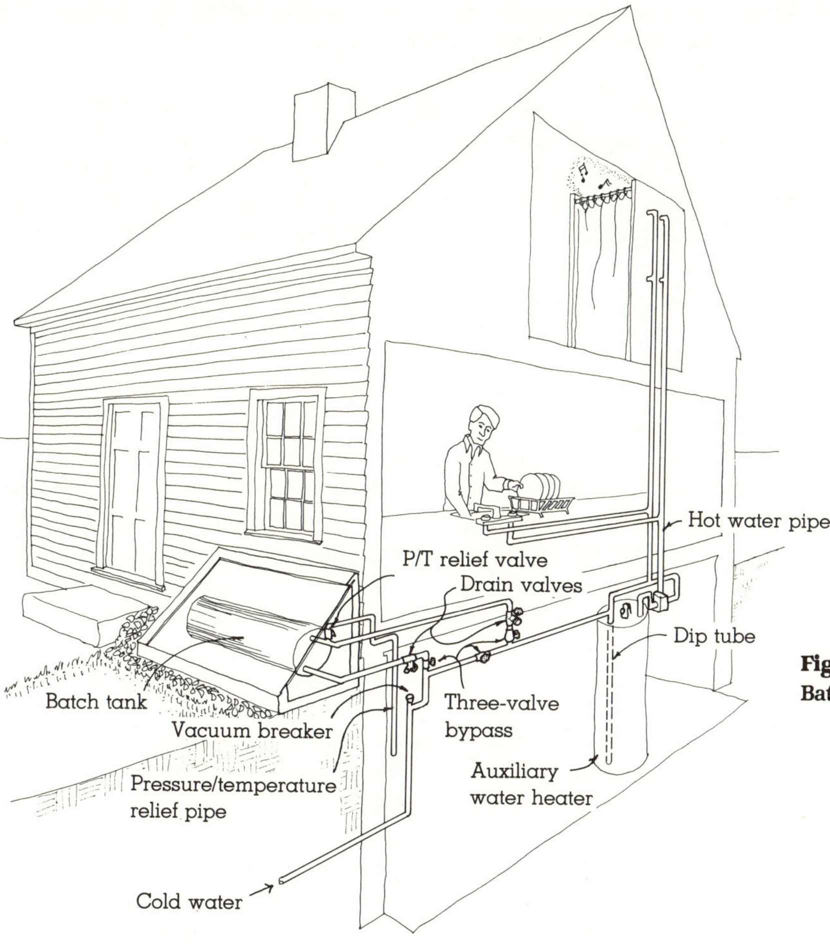

Figure 1–5.
Batch solar water heater system.

In a batch water heater, sunlight passes through the glass or plastic glazing and strikes the black water tank. The tank absorbs the sunlight, converting it to solar heat and warming the water in the tank. Reflectors shine additional sunlight onto the tank, thus increasing the solar heating. Insulation around the nonglazed areas of the tank keeps the water in the tank warm. This is especially important at night and on cloudy days. Only when you use hot water in your house does the warm water in the tank flow, either to the backup water heater or straight to the hot water faucet. The warm water from the tank is automatically replaced by cold water from your well or municipal supply. As long as the sun is shining, this new batch of water will be solar heated.

BUILDING YOUR OWN BATCH HEATER

This book is both a solar energy primer and a construction manual. First, it provides complete descriptions and analyses of batch solar water heaters from the first ones in this country in the 1800s to today's latest designs. Through examination of these designs you can determine which design is best for you. Both site-built and commercially available systems are presented. The book then provides everything you need to know to perform a thorough evaluation of the solar potential of your home, whether you live in a detached, single-family house or attached, multifamily housing. Fortunately, most homes are sited well for solar use and can benefit from solar water heating.

Step-by-step construction plans for two excellent batch solar water heaters are also given. You can follow these plans exactly, or use the construction details to guide you in building your own design.

Materials for a typical, one-family batch water heater cost between $50 for used materials and $800 for the very best new materials (1982 prices). With federal and state tax credits these costs can sometimes be reduced to less than half. Almost all of the materials needed are available at local lumberyards, hardware stores, and plumbing stores. For the few materials not easily available, mail order sources are provided. Analyses are also given of various types of glazing, water tanks, and other materials.

Depending on the design you choose to build and the plumbing layout of your house, you can completely construct your batch heater in one to five weekends. Only basic hand and power tools and moderate carpentry skills are needed. Since most do-it-yourself builders are more experienced with carpentry than plumbing, step-by-step instructions for soldering copper plumbing are included.

For homes that are not well sited for solar, other energy-saving water heaters beside solar should be considered (heat pump water heaters and heat recovery systems are possibilities). And, in all homes *conservation should be implemented as part of any energy-saving water-heating system.* Conservation alone can often save 30% of your water heating costs. Chapter 9 provides a complete analysis of this potential saving.

SOLAR HOUSE AND WATER HEATING IN COLD CLIMATES

Throughout many of the coldest parts of the United States and through much of Canada, passive solar house heating has become a reality. Many people have found passive solar to be both affordable and extremely effective. Solar greenhouses and large, south-facing windows with night insulation, only a few years ago the tools of solar researchers, are now becoming common. Thermosiphoning air panels and other passive solar heating systems are also being put to use. There is a fair chance that you may already have one of these passive systems.

The success of passive solar house heating can be attributed to its low initial cost, dependability, minimal maintenance, and effectiveness. On the other hand,

Figure 1–6. This batch heater is located in the upper space of a solar greenhouse. The removable exterior reflector boosts batch performance in spring and summer and is taken down in fall and winter. (Greenhouse and batch design by Dan Reif; greenhouse construction by Terry Fenton.)

active solar house heating, which uses separate components to collect, store, and distribute the solar heat and fans or pumps to move the solar heat through the system, has not become widely used. This is in part because the complexity of active solar has increased its initial cost while decreasing its reliability.

Passive batch solar water heaters have the same advantages over active solar water heaters. The modern history of solar house heating, which started with complex systems and gradually moved to simpler passive systems, is now being repeated. There is real beauty in things that work simply.

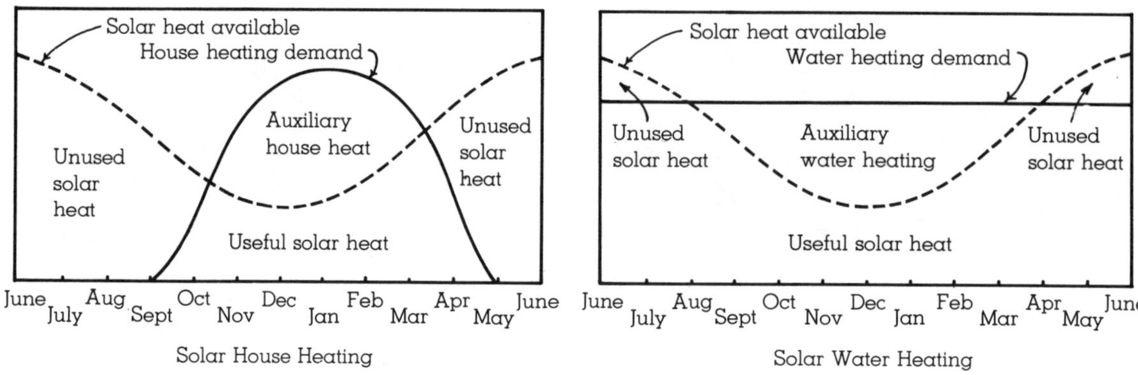

Figure 1–7. Solar water heating versus solar house heating: Solar water heating makes use of almost all of the available solar heat, whereas solar house heating only makes use of the available solar energy in winter and part of the spring and fall.

When you must decide between building a solar water heating system or a house heating system, the best choice of which to build first depends on several factors—two of the most important being the initial cost of the solar heating system and how much money can you save by using the system. Solar water heating and space heating systems generally operate within the same range of efficiencies. Therefore, the longer you can use the solar heating system during the year, the greater the potential for annual savings. This is why solar house heating is ideal in climates with long winters. A major advantage for solar water heating is that hot water is used fairly consistently throughout the year. If your batch water heater is designed to withstand freezing temperatures, or if freezing is not a problem in your climate, you can achieve this maximum solar utilization. If your solar water heating system is shut down in winter to protect it from freezing, then you lose some solar benefit.

Although the system may be shut down for several freezing months, the loss of total annual solar potential is often not that great. This is because during the winter the amount of solar potential is often significantly reduced by many cloudy days. For example, in a cold, cloudy environment where you shut down your solar water heater for three months (25% of the year) the loss may only be 15% of the annually available sunlight.

In cold climates the complete solar home of the early 1980s featured a passive solar room or greenhouse and an active solar water heater on the roof. This is about to change—passive solar will be used more frequently for heating water as well as houses.

CASE STUDY: UTILITIES AND BATCH WATER HEATERS

"Many regions in the United States are faced with seasonally imbalanced use of electricity. Typically, more electricity is demanded during the summer cooling season, resulting in high requirements for generation capacity which is underutilized during the winter season. Generation capacity is built to meet the peak summer load, which occurs during hot, summer afternoons. This 'simple approach' [batch] solar water heater performs best during the same general period.

*While there is little comparative data from which to draw definitive conclusions, it is fair to say that utilities faced with summer peaking problems should promote the use of 'simple approach' solar water heating as a small partial means of coping with summer peaking."**

Figure 1–8.
Union Electric Company's roof-mounted test batch collector. (Courtesy of Union Electric Company, St. Louis, Missouri).

This conclusion was reached by the Solar Hot Water Research Project sponsored by the Union Electric Company of St. Louis, Missouri. The primary goal of this project was to develop and test a low-cost solar water heater intended to operate between May and October. Because of the potential for freezing temperatures, the collector would be drained and shut down during the colder months. It was hoped that development and widespread use of solar water heaters would benefit utility customers with reduced water heating cost, and benefit the utility with reduced peak power demand. The project was started in early 1976. Charles

**A Simple Approach to Solar Heating of Domestic Hot Water,* Union Electric Company, St. Louis, Missouri, 1978.

Grandy served as project manager for Union Electric, with William Tao & Associates, Inc. as consulting engineers.

Several types of solar collectors were analyzed for their suitability, including initial costs and performance. It was determined that it would be best to use a solar water heater that would function without any external energy source other than water main pressure and would have no moving parts—a batch heater. High initial cost and potential maintenance problems forced Union Electric to reject active solar water heaters.

The first heater was constructed and installed on a test home in 1977 to determine its performance. Between May and October savings were approximately 40% of the total energy required to heat the domestic water.

An improved collector was tested during the same period in 1978. This second-generation collector was constructed by the home owner from plans developed by Union Electric. No problems were encountered during construction, demonstrating the feasibility of building these collectors as do-it-yourself projects.

Without solar heating, 5 to 6 gal. of 135°F water were produced for each kwh of electricity. The first collector saved sufficient energy to produce a seasonal performance of 8.6 gal. of hot water for each kwh of electricity used, and the modified design increased performance to 11.3 gal. of hot water per kwh—a 32% improvement over the first design.

The Union Electric designs have two 30-gal. water tanks, two layers of fiberglass glazing, no night insulation, and are painted flat black instead of using a selective surface (more on performance in chapter 3). With a moderate increase in the initial costs, better performance can be achieved, including use of the batch during colder weather.

Other data from tests of the second-generation Union Electric collector showed that there was as much as two-and-one-half times more hot water available with the solar heater than without it. Maximum water temperature in the collector was 180°F. The potential for producing such hot water resulted in notes being added to Union Electric's construction drawings advising home owners not to use water directly from the collector, but to use a thermostatically controlled mixing valve or to have the solar-heated water pass through the backup water heater to reduce potentially scalding temperatures. Maximum air temperature in this collector reached 158°F. This is important because very high air temperatures (over 190°F for long periods of time) can degrade the wood and other materials in the collector.

One of the major goals of this project was to reduce the peak demand for electricity. Peak demand in 1977 occurred at 4:00 P.M. on July 19. On this day from 2:45 P.M. through 7:45 P.M. the electric energy input to the electric water heater at the test home was *zero*. This occurred even though 51 gal. of hot water had been used during the 1:15 P.M. to 2:15 P.M. interval of the same day. From 2:00 P.M. to 2:45 P.M. the electric water heater used 2.5 kwh and then shut down until 7:45 P.M.,

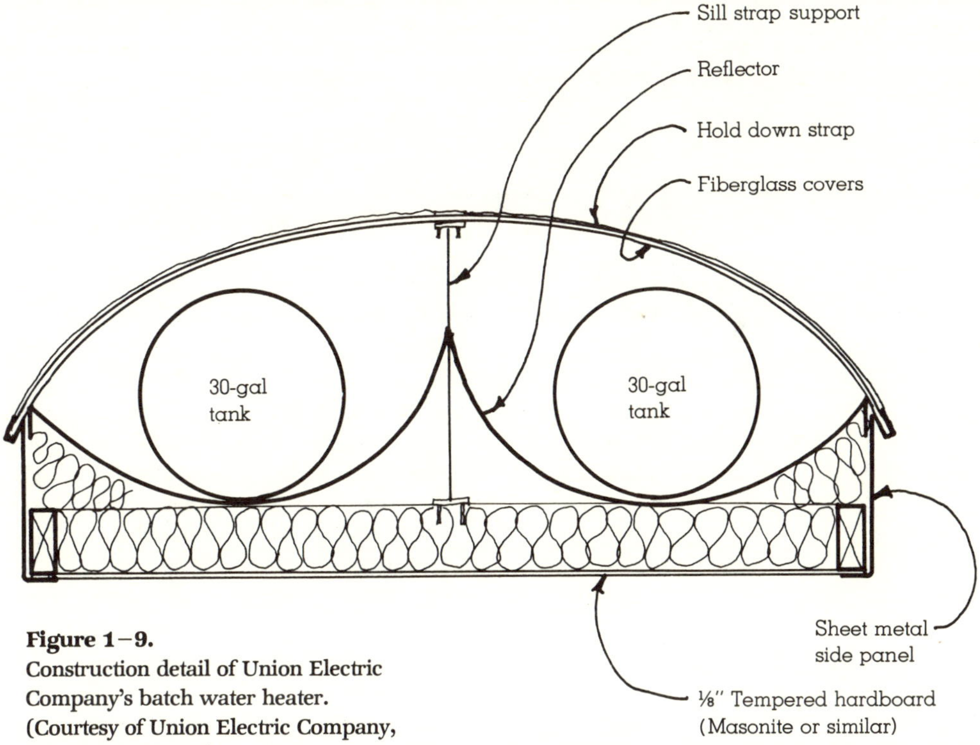

Figure 1–9.
Construction detail of Union Electric Company's batch water heater. (Courtesy of Union Electric Company, St. Louis, Missouri).

even though small amounts of hot water were used during this 5-hour period. The solar heater had supplied the remainder of the energy required between 1:15 P.M. and 7:45 P.M. to heat the water.

In 1978 the solar heater again provided enough hot water so that the backup water heater did not add to the peak demand, which occurred at 5:00 P.M. on August 24. In fact, the water heater did not use any electrical energy from noon through midnight on this day even though 28 gal. of hot water were used during this time.

Testing of these solar collectors was expanded to include twelve additional homes during the summers of 1979 and 1980. Proving the soundness of the design of these collectors, there have not been any operational or maintenance problems with any of the fourteen collectors over the two-year testing period. Projections of the total seasonal energy savings attributed to the solar water heaters are shown below.

Solar Water Heater—Projected Energy Savings and Performance In St. Louis, May 15 to October 15

Residence	*Number in Family*	*Type of Water Heater*	*Solar Contribution*			
			1979		*1980*	
			Btu	*%*	*Btu*	*%*
1	3	Electric	1,961,676	42	data not available	-
2	3	Electric	2,032,272	46	1,467,872	42
3	6	Electric	2,812,784	31	1,370,720	30
4	4	Gas	2,404,665	22	2,152,376	15
5	5	Gas	6,424,920	36	6,326,978	35
6	6	Gas	5,468,822	49	4,895,582	54
7	4	Gas	303,437	7	2,968,230	47
8	3	Gas	5,191,100	33	3,658,263	15
9	3	Electric	data not available	-	data not available	-
10	4	Gas	2,466,090	29	2,802,348	27
11	7	Gas	1,674,422	20	data not available	-
12	4	Gas	1,374,635	27	2,634,100	32
13	3	Gas	1,639,834	39	1,254,616	28
14	4	Gas	2,125,166	33	3,296,640	40

2 BATCH WATER HEATERS: PAST AND PRESENT

Figure 2–1.
The absolute basic batch today. (Courtesy Basic Designs, Inc.)

Batch water heaters have had a long history in the United States. The first recorded batches in this country appeared during the 1800s on farms and ranches. These designs consisted of the absolute basics: a black can of water in the sun. This simple, functional design is still in use today with surprisingly good results.

HISTORY OF BATCH HEATERS

The first commercial solar water heater in the United States was patented in 1891 by Clarence M. Kemp. Mr. Kemp's Climax Solar Water Heater combined the black-painted tank in the sun with the principle of a "hot box." His basic system consisted of four 8-gal. heavy galvanized metal tanks enclosed in a pine box insulated with felt paper and glazed with a single layer of glass. The tanks were painted a flat black and plumbed in series with metal tubing. The systems were mounted on either the roof or south wall. This batch was sold in California as an alternative to the wood stove water heater and early gas water heaters. The wood stove heater was especially undesirable during the warm weather, and gas for the heaters was expensive.

Performance of the Climax was very good for its simplicity. As the promotional literature stated: "During the early Spring and late Fall months, while the temperature during the day has been near the freezing point, the heater water has been over 100 degrees, the efficiency shown by the heater is from 60 to 100 degrees greater than the temperature for the day. The water at times almost boils."

By 1900 more than 1,600 Climax heaters had been installed in California. It was recalled that "Everyone had one. There was nothing uncommon about it at all. I can't remember a house on the block that was built at that time or soon after that didn't have a solar heater."*

There were eight models of Climax heaters ranging in size from 32- to 700-gal. capacity. The price of the smallest and most popular one was $25. At this price it took less than three years of energy savings for the heater to pay for itself.

In 1898 Frank Walker improved Kemp's Climax design and applied for a patent on the "Walker Combined Solar and Artificial Heat Water Heater." Walker's heater had one or two 30-gal. tanks that were set into the roof so that the glass glazing was flush with the roof. By placing the tanks in the roof, heat loss was reduced and the system had clean visual lines. The reduced heat loss allowed Walker's design to have warmer water in the morning.

Another major feature of the Walker Solar Heater was that it was connected directly to the backup heater. This made it the first solar water preheater.

In 1905 Charles Haskell of the Solar Heater Company constructed a batch, which he called the Improved Climax. His design used a shallow rectangular tank instead of the cylindrical tank previously used. This modification provided a larger collector surface area in relation to the volume of water. The rectangular tank of water heated faster and was warmer on cloudy days. In warm climates the Improved Climax, along with the Climax and Walker, supplied most of the domestic hot water needs for eight or nine months of the year.

In 1909 William Bailey developed and started selling thermosiphoning solar water heaters. He called his design the Day and Night solar water heater. Initial work on the thermosiphoning principle had been done by Bailey's physician, Dr. Remington, who separated the solar water heater into two units: a solar heat collector and a water storage tank. The collector consisted of a coiled pipe placed inside a glass-covered box that was mounted outside the house facing south. The small volume of water inside the pipe was quickly solar heated, which caused it to

---

*Walter van Rossem, son of a former Climax owner, from Ken Butti and John Perlin, *A Golden Thread: 2500 Years of Solar Architecture and Technology.*

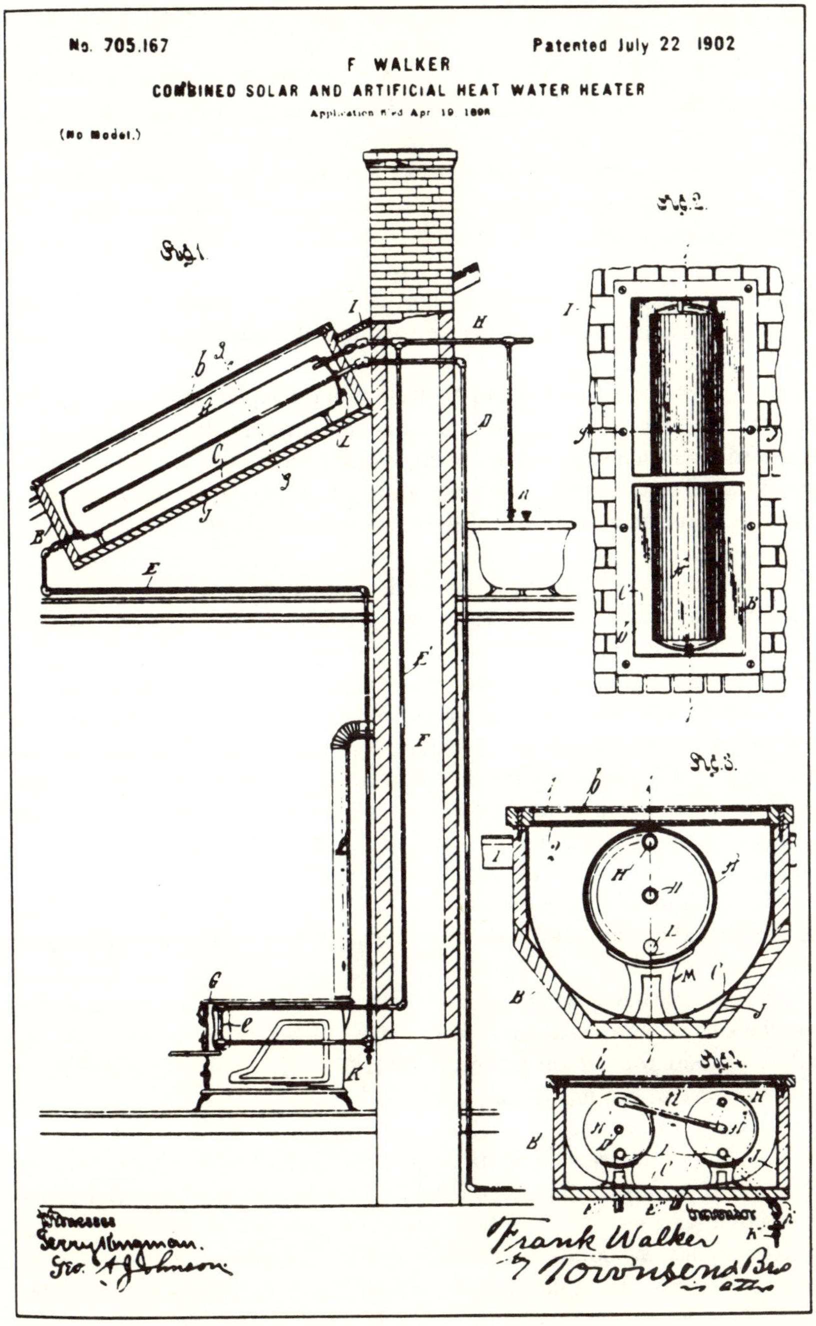

Figure 2–2. Patent drawing for the Walker Combined Solar and Artificial Heat Water Heater, 1902, the first batch preheater.

rise and flow into the tank inside the house. Cool water in the tank flowed into the bottom of the pipe to replace the warmed water. Bailey improved Remington's design by insulating the water tank with 3'' to 9'' of powdered limestone. Bailey guaranteed that the water temperature would not decrease by more than 1°F per hour. Although the cost of the Day and Night was considerably more expensive than the simple batches, its performance was unmatched. The improved performance and strong promotion of the Day and Night eventually led to the decline of the batch water heaters.

The discovery of abundant supplies of natural gas and oil, and strong promotion for these energy sources, led to the decline of the entire solar water heater industry in California. By 1930 the once flourishing solar industry there had been severely diminished.

In Florida solar continued to compete well with the high cost of conventional fuels. By 1941, 60,000 systems had been installed. More than half the population of Miami used solar-heated water. But, during World War II the solar industry was temporarily halted by the freeze on the use of copper. After the war the availability of inexpensive, government-subsidized electricity eventually faded the solar industry there too.

During the 1950s and 1960s the Japanese developed an extensive assortment of passive solar water heaters. Their designs ranged from simple unglazed plastic bag collectors to self-contained thermosiphoning systems.

One of the simplest designs is the solar pillow. Hundreds of thousands of low-cost solar pillows have been used throughout Japan. At the peak of their sales they were even available in many urban department stores. The pillow consists of a clear or black plastic top layer and black plastic bottom. An additional clear plastic layer could be added on top in the winter. The collector is filled each morning with 40 to 50 gal. of water, then emptied at night. This complete system sold for $6 to $10, making it accessible to almost everyone. The pillow lasted about two years, but the savings far outweighed the small cost. But by the late 1960s the availability of cheap energy and the desire for more convenience brought a sharp drop in sales of solar pillows.

Another Japanese batch system uses several cylinders set into an insulated box similar to the early Climax heaters. During the mid 1960s, sales of these systems averaged more than 1/4 million per year. By 1969 more than 3.7 million solar water heaters were in use in Japan!

The Hitachi Hi Heater and Sekisui Solar are examples of the current multitank systems. The Hitachi Hi Heater uses six polyethylene cylinders that hold a total of 44 gal. of water. The tanks are enclosed in an insulated box and have a single layer of polycarbonate glazing.

Batch design has also flourished in other parts of the world. One Russian batch design is a variation on the Japanese solar pillow. The Russians use glass divider strips to create a double glazing effect and to direct the water flow.

Figure 2–3. Steve Baer's Breadbox Water Heater.

In Australia, the Suntrap uses a single 10-gal. tank set in an aluminum enclosure that is single glazed. In New Zealand, the SAV Solar System uses double glazing around a 10-gal. tank with a large reflector behind it. Both of these heaters are now available in the U.S. The Suntrap has been somewhat modified to meet the greater demand for hot water in this country and is marketed here as the Sunflow water heater.

With the 1973 oil embargo, American interest in solar resumed. Steve Baer of Zomeworks in Albuquerque, New Mexico, developed an innovative batch heater that has become a highlight of modern batch design. His design consists of two horizontal tanks in a double-glazed rectangular box whose lid and front can be opened as reflectors during the day and closed as insulation at night. These reflective/insulating shutters increase solar heat gain during the day, but also significantly reduce heat loss at night. Thus, this design improvement extends the range of climates where the batch heater can be used.

John Hammond, a fellow solar enthusiast, noticed the similarity between Steve Baer's design and a breadbox. Since then Baer's design has been known as the Breadbox Water Heater. Breadboxes have been built throughout the country from plans sold by Zomeworks.

SITE-BUILT BATCH SYSTEMS

Site-built batches offer the owner/builder the greatest flexibility of design and cost. The system can be one of a kind or standardized and built from plans. The following site-built batches are offered to show the variety of systems possible. Plans are available for many of these systems; ordering information is given on page 37.

Figure 2–4.
Nature's Way Energy Systems offers a batch module, installed here in unusable attic space. The system consists of one 45-gal. tank with 46″ x 96″ low-iron glass outer glazing and two pieces of 17″ x 88″ glass inner glazing. An electric heating element in the batch tank provides auxiliary water heating during the warm months when the backup heater is bypassed. (Courtesy of Nature's Way Energy Systems.)

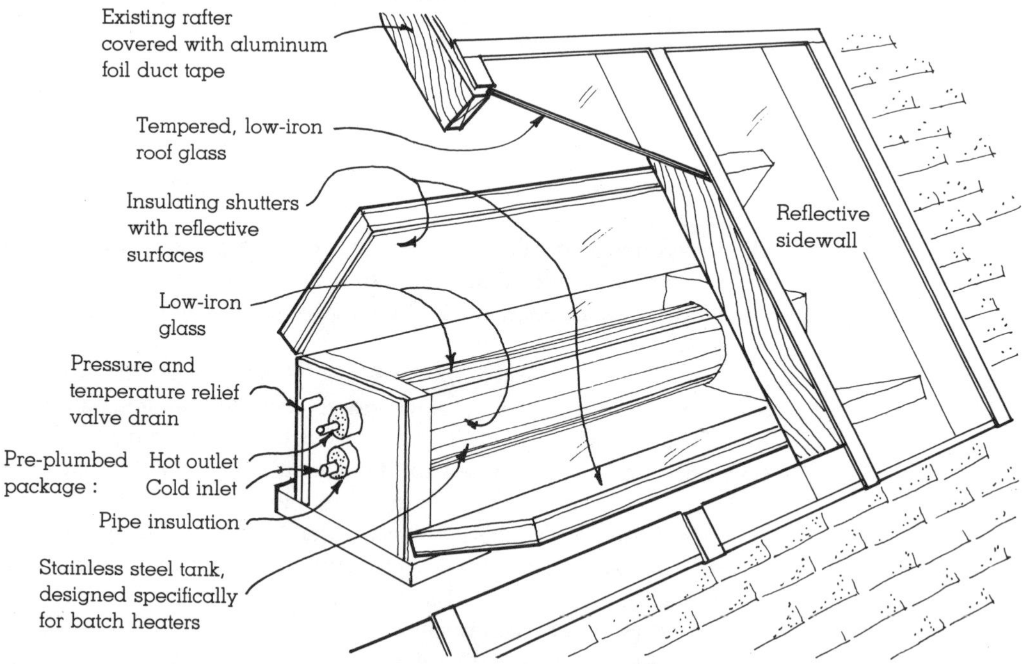

Figure 2–5. Nature's Way Energy Systems' Batch Module mounted in an attic. Among other advantages, the insulating shutters allow the effective use of an electric heating element in the batch tank for backup heating.

Nature's Way Energy Systems

Nature's Way Energy Systems in Keene, New Hampshire, is one of the leaders in the development and widespread use of batch systems in New England. Len Meserve and company conduct batch construction workshops, consult on designs with builders and home owners, and install systems. Their recent projects include a workshop in which high school students built a batch for their shop class, design and construction of a batch system on each new house in a forty-one house subdivision, and development of a batch module. The module is available as a kit ($950) or completely assembled ($1,075 retail), which saves much on-site construction time.

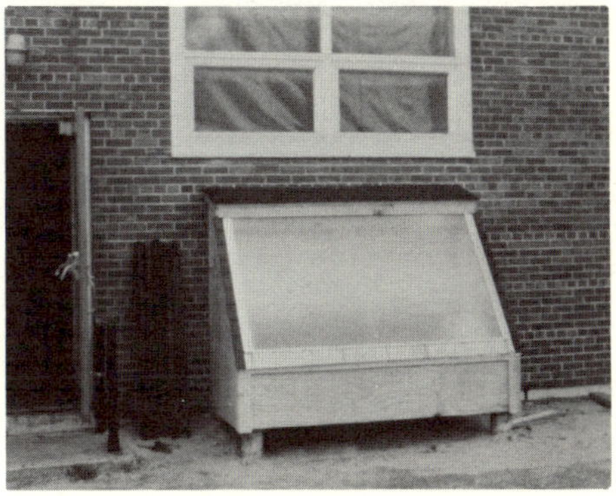

Figure 2–6. Batch system constructed by a high school workshop. (Courtesy of Nature's Way Energy Systems.)

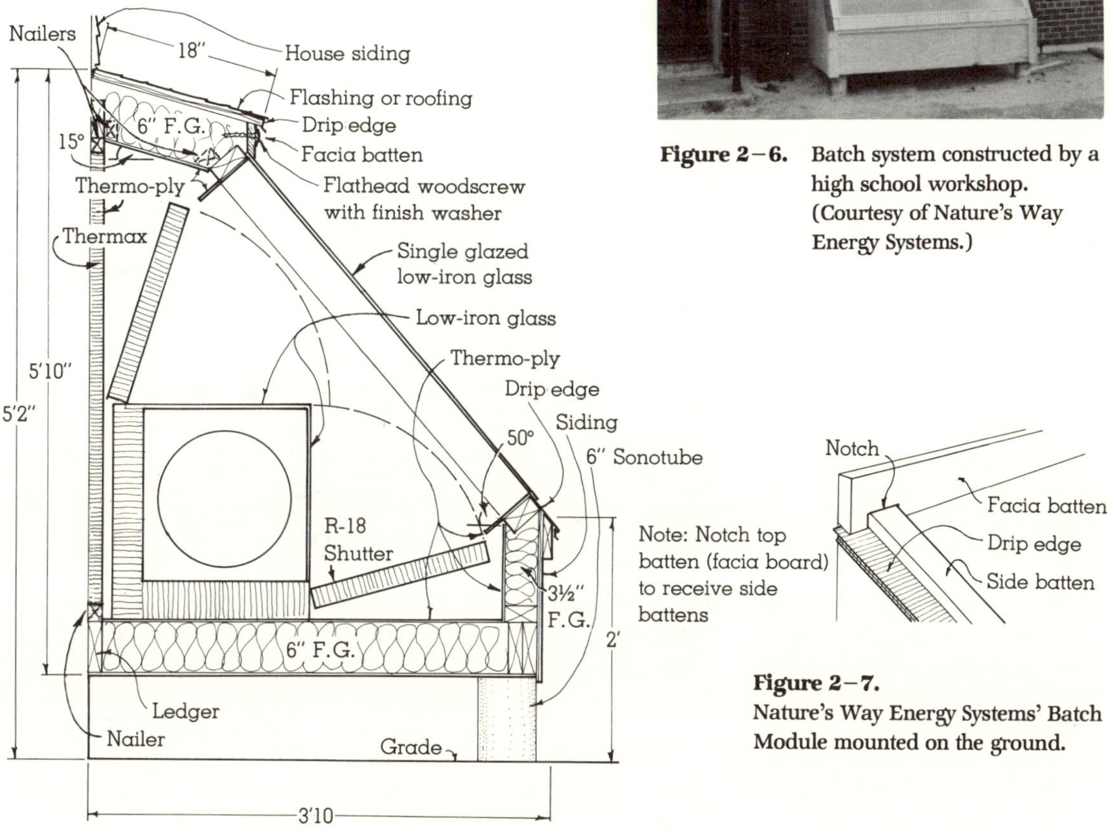

Figure 2–7. Nature's Way Energy Systems' Batch Module mounted on the ground.

Figure 2–8.
Batch in attic of solar greenhouse in subdivision in Keene, New Hampshire. (Courtesy of Nature's Way Energy Systems.)

Figure 2–9.
This attic batch is drained for the winter and a 150 CFM (cubic feet per minute) fan is used to move solar-heated air from the collector to the living room—an option greatly appreciated by the home owners. (Courtesy of Nature's Way Energy Systems.)

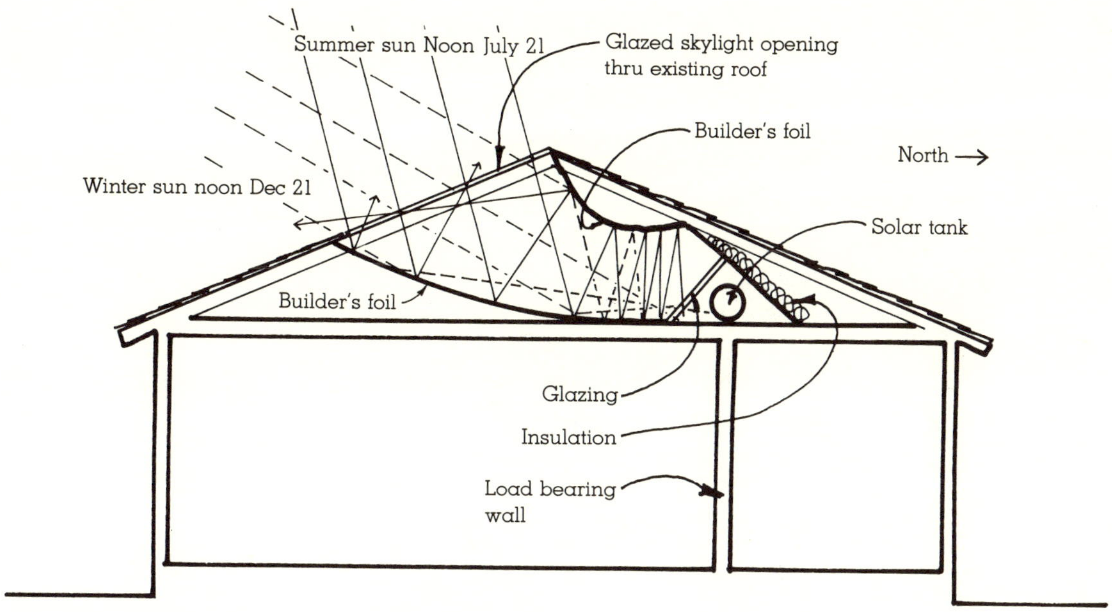

Figure 2–10. John Golder's Foil Funnel Skylight/Batch.

Foil Funnel Skylight/Batch Heater

A proposed batch design with a great deal of promise is John Golder's Foil Funnel Skylight/Batch. A funnel made of builder's aluminum foil is used to direct sunlight coming through a south-facing skylight to a solar tank located directly over a bearing wall of the house.

Golder has already used this technique of the foil funnel to direct southern light into non-south-oriented rooms. It can provide passive solar heat to rooms that would otherwise be difficult to solarize, as well as possibly make the skylight qualify for some tax credits.

The roof glazing can be several times larger than the skylight or solar water tank, effectively concentrating the solar heat. Besides replacing the traditional framed light well with inexpensive foil, this technique requires no reframing of roof rafters and ceiling joists. Simply wrap all wood members within the funnel with foil. The flexibility of this system makes it ideal for batch heater retrofits.

Solstice Designs

Malcolm Lillywhite of Solstice Designs, Inc. designed a batch water heater while working in the remote villages of Lesotho and Botswana, Africa. Collector enclosures have been constructed from mud and dung, wood, brick, or almost any available construction material. A grass mat is used for night insulation. Only the glass or fiberglass-reinforced glazing has to be imported. It is a very low-cost

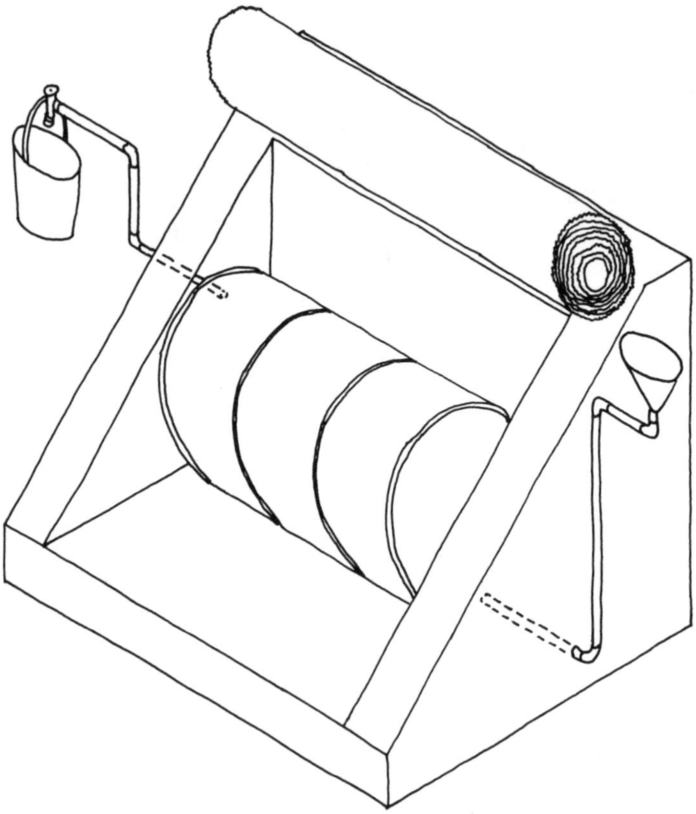

Figure 2–11.
Solstice Designs batch requires the user to add a bucket of cold water at one end to receive hot water at the other end.

system. To use the collector requires adding fresh cold water to the tank, which is replaced at the other end with hot water.

Solstice has also developed a horizontal tank(s) batch water heater, as well as several thermosiphoning and active solar water heaters. All five plans are available from them for $24.50.

Greenhouse Batch

Solar designer and builder Tom O'Brien constructed the sheltered batch shown in figure 2–12 in Aspen, Colorado. Two 30-gal. tanks with a selective surface on their underside are neatly tucked into the top of a solar greenhouse. A manually operated reflector/insulating shutter can be adjusted to maximize solar gain to the tanks throughout the year, and at night covers the outer glazing as well as allows warm greenhouse air to circulate up to the collector. The 35 sq. ft. of double glazing provides a ratio of 1.7 gal. of water per square foot.

Figure 2–12.
The tanks for this batch heater designed by Tom O'Brien are tucked under a well-insulated greenhouse roof.

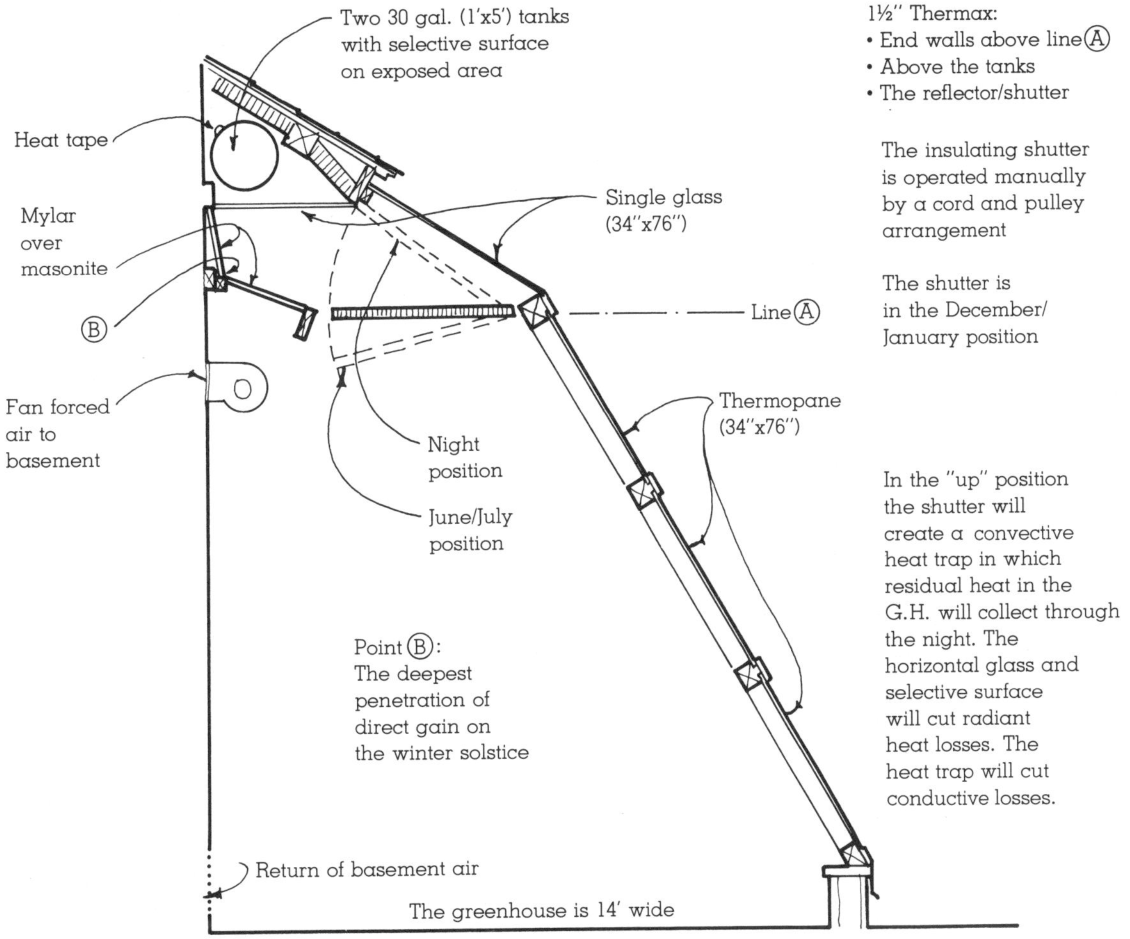

Figure 2–13. Tom O'Brien's greenhouse with batch heater.

Figure 2–14.
Marc Rosenbaum's batch retrofit to a solar greenhouse.

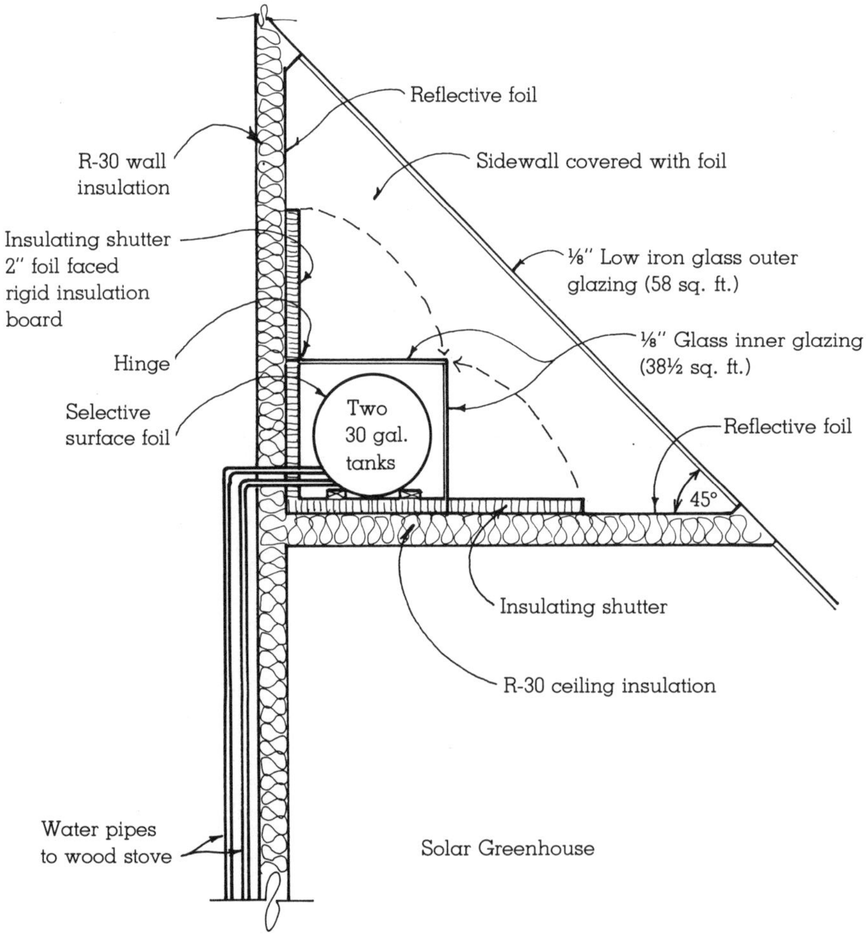

Figure 2–15.
The high-performance batch designed by Marc Rosenbaum makes effective use of its reflector/insulating shutter.

Batch Retrofit to Solar Greenhouse

Marc Rosenbaum of Energysmiths in Meriden, New Hampshire, added a high-performance batch to his solar greenhouse. This addition was not originally planned. Because backup water heating is provided by a thermosiphoning system from the wood stove and his small, superinsulated solar house rarely needs heat, Rosenbaum wanted to eliminate the use of the stove for water heating in all but the coldest, cloudiest weather. The collector contains two 30-gal. tanks covered with Maxorb selective surface foil, two layers of glass, and a reflector/insulating shutter. The glazing is oriented directly solar south and is sloped at a 45° angle. (The latitude of the site is 43°.)

The shutter is located in a large space between the glazings. This keeps the reflective surfaces shiny and free of snow and ice. The outer glazing is approximately one and one-half times larger than the inner glazing, which maximizes solar heat gain. The inner glazing and close-fitting insulating shutter are kept as small as possible to minimize heat loss. (This is an example of solar researcher Dihn Khanh's performance-boosting concept of building a large inexpensive "slave" collector to house a smaller high-price collector. Solar researcher William Shurcliff analyzes this concept and ninety-nine others in his book *New Inventions in Low-Cost Solar Heating,* listed in the bibliography.)

With the outer glazing providing almost one square foot of glazing for each gallon of water, performance has been excellent. In summer water temperatures range from 90°F to boiling! Clear-day winter water temperatures are between 80°F and 120°F. On cloudy winter days and at night the insulating shutter is closed and the wood stove heats the water. The total system cost including the thermosiphoning wood stove connection was $800.

Integral Design

John Burton of Integral Design in Sacramento, California, has been involved in the construction of more than thirty batch heaters of various designs during the last few years. Figure 2–16 shows one of his collectors with two 42-gal. tanks, two layers of fiberglass glazing, and an external reflector to increase winter gain. Total material costs for this system in 1978 were $400: $150 for plumbing, $50 for glazing, and $200 for lumber and supplies for the collector box. The recycled tanks were free. Burton no longer uses the small, fixed reflector shown in this design because it was difficult to construct and added little to the heater's performance. When a reflector is desired he recommends an operable reflector/insulating shutter like the one shown in figure 2–19.

The performance of this system in California has been excellent. From May to October the batch provides 100% solar-heated water. Annually, it provides 70% of the hot water needs of a family of three. Of course, Sacramento's climate helps with winter temperatures in the 50s and 60s during the day and 30s and 40s at night, summer temperatures usually above 90°F, and 79% annual sunshine.

Figure 2–16.
Batch with small overhead reflector designed by John Burton of Integral Design.

Figures 2–17, 2–18, and 2–19 show batch heaters that were constructed during Burton's workshops. Figure 2–17 shows a collector in which two 42-gal. tanks are located directly over the house's exterior bearing wall. With two layers of fiberglass glazing and flat black tanks, in Sacramento the system is expected to provide 40% of the annual hot water demand for a family of four.

Figure 2–19 shows a similar collector with an additional reflector/insulating shutter. The reflector adds a significant amount of solar gain during the day, but does not seal tight enough around the glazing to significantly reduce night heat loss. This system is currently being monitored by the Pacific Gas and Electric Company.

Figure 2–17.
One of the batch heaters constructed during an Integral Design workshop.

Figure 2–18.
Integral Design workshop batch located over the house's exterior bearing wall.

Figure 2–19.
Batch with reflector/insulating shutter, also built during an Integral Design workshop.

Figure 2–20. John Golder's Capsule Collector with double reflectors.

Capsule Collector

John Golder of Sunny Side Up in Santa Cruz, California, has been building his Capsule Collector since 1978. This design emphasizes low cost and ease of construction. The Capsule Collector features one or two horizontally mounted 30-gal. tanks and a manually operated wrap-around reflector/insulating shutter. The collector is double glazed using fiberglass-reinforced plastic outer glazing and Teflon or Tedlar inner glazing. Several layers of glazing tape are used to separate the glazings, thus providing insulating air spaces between them. The reflector/insulating shutter is constructed of heavy waterproof fabric (e.g., nylon-reinforced vinyl or canvas) for the outside, 3 1/2'' of fiberglass insulation, and galvanized sheet metal (24 gauge or lighter) with reflective material (aluminized Mylar sold as the space blanket) on the inside. An additional reflector (as shown in figure 2–20) can be placed just below the tank. On a sunny day each Capsule Collector tank can raise the water temperature by 30° to 50°F. With mild night temperatures, the water temperature typically drops 10° to 20°F.

Figure 2–21. Crystal City batch heater.

Crystal City

In 1980 the National Center for Appropriate Technology in Butte, Montana, funded a demonstration project under which 120 batch water heaters were fabricated and installed in Crystal City and six surrounding towns in southern Texas. The Center for Maximum Potential Building Systems in Austin, Texas, designed the collectors to be low cost (using recycled and locally available materials) and easy to construct. This collector is unique in its integrated design. It is sized to optimize material dimensions and can be made of used or new materials—or a combination of both. The recylced collector even utilizes cleared out fluorescent light tube glazing. Using all recycled materials, the collector cost can be as low as $18 per unit; with new materials, approximately $140. Due to the dimensional coordination effort, the collector is easy to construct on a minifactory basis using low-cost, readily available equipment. The Crystal City factory cost about $4,000 (not including the cost of the building), and is designed to produce four collectors per day using five laborers. There are now more than 400 collectors based on the Crystal City design operating in South Texas. Because these collectors are designed for the warm climate of southern Texas, they should be modified when used in colder climates.

This simple batch system uses a single 30-gal. tank that is painted flat black and surrounded by a single layer of fiberglass-reinforced plastic. Due to the unique design, the reflector enables a wider range of orientation (30° east or west of south), which is an immense help in retrofitting a town by following existing roof angles. The reflective surface is made of either recycled aluminum printing plates,

Figure 2–22.
Internal cusp reflector batch by Rodale. (Courtesy of Rodale's *New Shelter* magazine, copyright © 1981 Rodale Press, Inc. All rights reserved.)

Figure 2–24.
Lita Furby's batch with cusp reflector. (Courtesy of David Bainbridge.)

aluminum truck siding, new anodized aluminum, or white painted wood. The reflector's rigid wood frame acts also to distribute the weight of the tank, allowing the collector to be supported by the structure of even relatively poor roofs (assuming the roof can support at least 15 lb./sq. ft).

In southern Texas this collector supplies 40 gal. of 105°F water, 300 days a year. Performance of this collector with a white painted wood reflector was only slightly lower.

Plans for this batch collector and collector factory are available from the Center for Maximum Potential Building Systems.

Rodale's Batch Heater

Rodale's *New Shelter* magazine batch heater uses a single 40-gal. vertically inclined tank with selective surface foil covering, triple glazing, and an internal cusp reflector. The glazing consists of an outer layer of fiberglass-reinforced plastic and two inner layers of Teflon. The reflector is made from aluminized Mylar glued to 1/8″ hardboard, which is bent into a cusp shape. Detailed, step-by-step construction plans are provided in the July/August 1981 issue of *New Shelter*. The cost of this batch before tax credits is $750, which includes everything from insulating the home's existing pipes to building and installing the solar water heater.

Figure 2–23.
Batch frame and rear of cusp reflector. (Courtesy of Rodale's *New Shelter* magazine, copyright © 1981 Rodale Press, Inc. All rights reserved.)

Figure 2–25.
Rael and Carl Rosato's vertical tank batch.

Cusp Reflector Batch

Lita Furby constructed this system on her home in Eugene, Oregon. It consists of two 30-gal. tanks with selective surface foil, internal cusp reflector, and triple glazing. The roof angle and structure made a tilted mount necessary. She tested the collector's performance with a black interior and later with the cusp reflector. Both designs performed equally well, providing full hot water into October. Auxiliary water heating is supplied by thermosiphoning from a wood stove heat exchanger.

Year-Round Batch

Rael and Carl Rosato constructed this batch in Oroville, California, to provide all of their hot water. (Previously, they heated their water on the stove.) With lots of sun guaranteeing hot water in the summer, they used steeply sloped glazing to maximize winter performance. To simplify construction they placed the tank vertically. Because the collector has 1 sq. ft. of glazing for each gallon of water, the water gets hot very quickly. On sunny winter days, water temperature averages well over 100°F. With no auxiliary water heater, the owner/builders are very satisfied.

Figure 2–26.
Triple-tank batch designed by Dan Plambeck. (Courtesy of David Bainbridge.)

Triple-Tank Batch

Dan Plambeck constructed this triple-tank batch using selective surface foil on the top front quarter of the tanks and double fiberglass-reinforced plastic glazing. The interior of the collector and the rest of the tanks are painted flat black. The pipes to the house are located alongside the gas heater duct to provide additional heating and freeze protection. The horizontal layout was selected to bridge the main roof beams and to minimize visibility from the street.

NCAT

The National Center for Appropriate Technology (NCAT) distributes plans for two batch designs: a low-cost, single-tank horizontal system designed by the Community Action Agency in Santa Barbara, California, and a two-tank horizontal system designed by the NCAT staff.

The low-cost Santa Barbara design uses a single layer of fiberglass glazing with an insulating blanket for additional nighttime heat retention. This design is not recommended for cold climates.

NCAT's design can be used in northern climates. It combines double fiberglass glazing and hinged insulating shutters. The two tanks are mounted horizontally one above the other, and are plumbed in series. Because of its weight, NCAT recommends that this collector be mounted on a foundation on the ground, not on a roof.

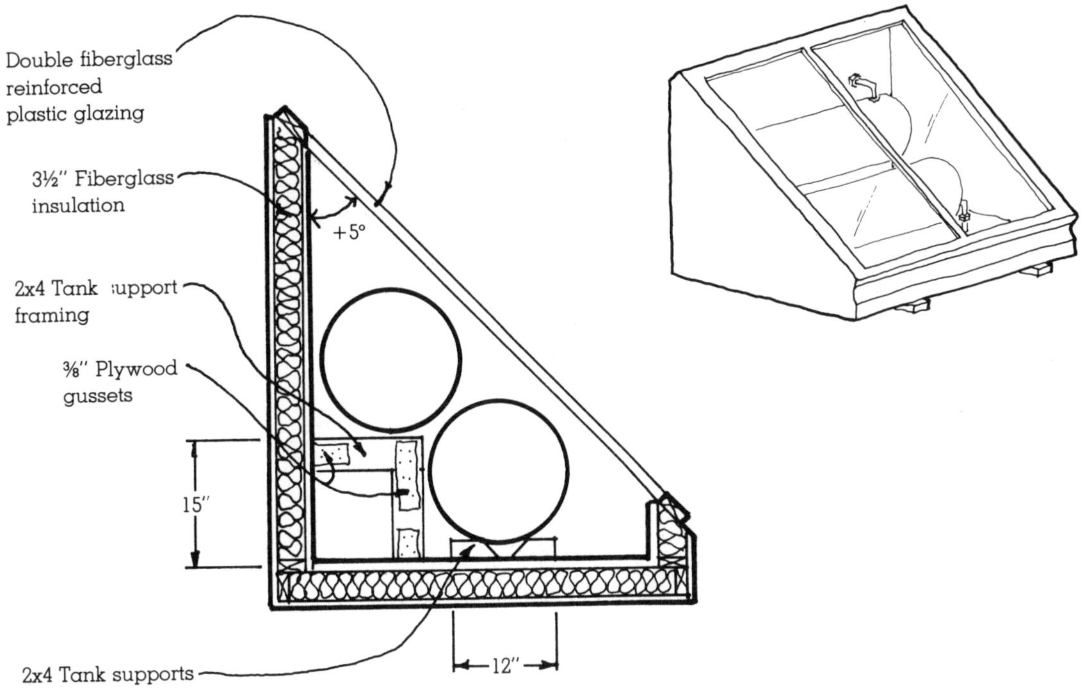

Figure 2–27. Batch design by the National Center for Appropriate Technology.

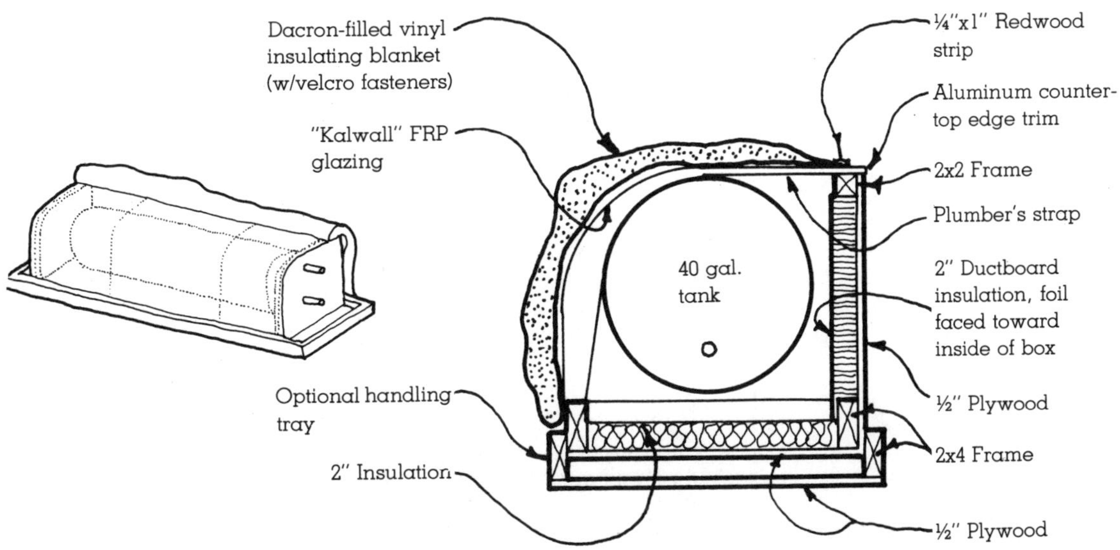

Figure 2–28. Batch design by the Community Action Agency in Santa Barbara, California.

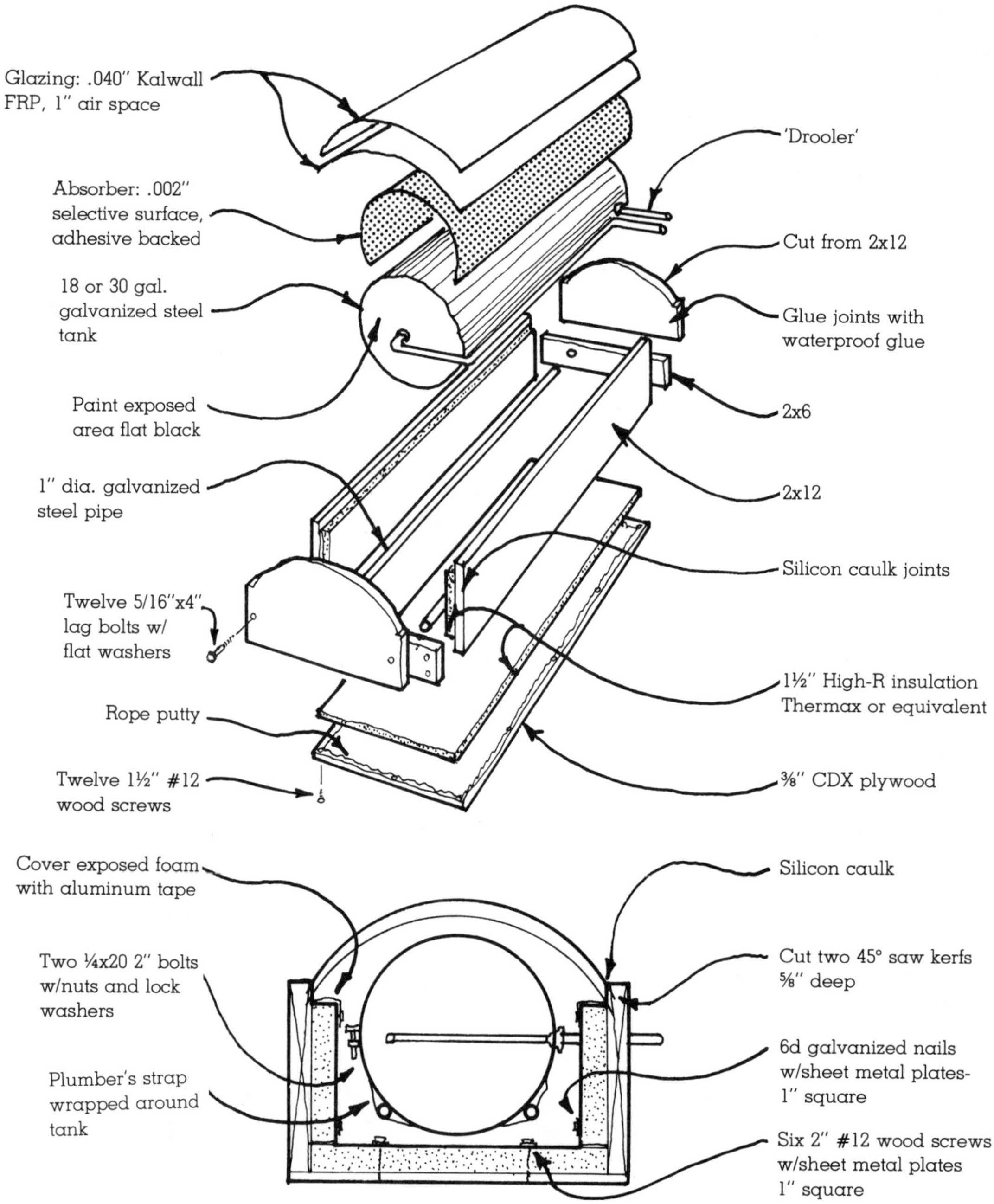

Figure 2–29. Sol-Airesearch batch.

Figure 2–30.
Conservation Concepts' Sundrop module.

Sol-Airesearch

Sol-Airesearch of Branford, Connecticut, used a DOE Appropriate Technology grant to develop a batch incorporating a single tank covered with a selective surface and double fiberglass-reinforced glazing. The performance of this batch has been monitored, and the results are discussed in chapter 8.

Sundrop Batch Kit

Conservation Concepts, Inc. of Hummelstown, Pennsylvania, has been involved in batch design, construction, and hands-on workshops since early 1980. Their latest design is the Sundrop, a modular approach in which each tank has its own enclosure, rather than two tanks in one larger box. Two or more Sundrops are used in series where additional hot water is needed. The Sundrop is available as a precut kit (price not yet available) or as a ready-to-install unit ($975). The unit can be installed on the ground, wall, or roof, or behind a skylight.

This is a high-quality design featuring a 46-gal. stainless steel tank, copper plumbing, selective surface, and triple glazing. The stainless steel tank is 18 gauge with 3/16'' thick ends. Four fittings are used at one end for the hot water outlet, pressure/temperature relief valve, optional temperature sensor, and cold water inlet. The hot water outlet and cold water inlet include diffuser tubes that run the length of the tank with evenly spaced holes to encourage gentle water movement through the tank during periods of use. This maximizes temperature stratification. A selective surface foil is bonded only to the sunny side of the tank. The triple glazing consists of two layers of 46'' x 76'' tempered glass and a single inner layer of Teflon. This assembly is enclosed in a wood frame with a plywood skin. Two inches of isocyanurate foam with reflective foil surface are used to provide R-16 insulation and internal reflectors.

Figure 2–31.
Batch offered in kit form from Solar Components.

This collector is not prone to freezing, except in very cold weather. To protect the pipes from freezing, Conservation Concepts recommends that the Sundrop be integrated into a nonfreezing space, warm house air be allowed to circulate into the collector, the collector be drained during the coldest months, or electric heat tape be used.

One 46-gal. Sundrop is sized to provide 40% to 60% of the annual hot water needs of a household of two or three. Larger households should consider two units. Summertime temperatures can be expected to exceed 160°F, while 80°F to 120°F temperatures are typical for sunny winter days.

Further information can be obtained from Conservation Concepts, Inc., 484 Middletown Road, Hummelstown, PA 17036.

Solar Components Batch Kit

Solar Components Corporation markets a batch water heater kit and factory-assembled unit. Their collector consists of two flat-black, 30-gal. glass-lined steel tanks, two layers of Sun-Lite fiberglass glazing, rigid foam insulation, and optional electrical freeze protection. The collector can be easily ground mounted with an optional mounting rack ($89). Annual savings from use of this batch heater is expected to exceed 6 million Btu.

The freeze protection consists of two thermostatically controlled silicone pad heaters factory mounted to each tank. Heat tape is recommended to protect pipes from freezing.

The kit price is $414 (plus $20 for shipping) and $80 for optional tank freeze protection. Factory assembled the collector costs $1,295 ($1,395 with freeze protection). To order, write to Solar Components Corp., P.O. Box 237, Manchester, NH 03105.

Batch Plans

Center for Maximum Building Potential, Crystal City Collector, 8604 Webberville Road, Austin, TX 78729; one tank, vertically inclined with exterior reflector.

Farallones Institute, Rural Center, 15290 Coleman Valley Road, Occidental, CA 95465 (Attention: Peter Zweig).

John Golder, P.O. Box 854, Santa Cruz, CA 95061; one horizontal tank with reflective insulating cover; $6.

Handyman magazine, "Build a Solar Water Preheater," B. Speer, January 1981.

Integral Design, 4708 Raley Boulevard, Sacramento, CA 95838; horizontal and vertical two-tank design; $15.

National Center for Appropriate Technology (NCAT), P.O. Box 3838, Butte, MT 59702; one and two horizontal tanks; low-cost systems.

Nature's Way Energy Systems, Route 32, Keene, NH 03431; Batch Module with one horizontal tank and reflector/insulating shutters; $10.

New Shelter magazine, "Sun on Tap—The Best We Know," by F. Langa, July/August 1981; one-tank, vertically inclined batch with cusp reflector.

Passive Solar Institute, P.O. Box 722, Davis, CA 95616; three tanks, vertically inclined; $10.

Sol-Airesearch, Inc., 305 Thimble Islands Road, Branford, CT 06405; one tank vertically inclined; plans free with stamped, self-addressed envelope.

Solar Components Corp., P.O. Box 237, Manchester, NH 03105; two tanks, vertically inclined.

Solstice Publications, Box 2043, Evergreen, CO 80439; one horizontal tank design plus four other solar water heaters; $24.50.

Union Electric—Solar Preheater, P.O. Box 149, St. Louis, MO 63166; two tanks, vertically inclined, low-cost batch system.

Zomeworks, P.O. Box 712, Albuquerque, NM 87103; Breadbox (horizontal tank with reflector/insulating shutters), and Pro (one vertical tank with slip-over night insulation); $6.

COMMERCIAL BATCH HEATERS

Commercial batch heaters are becoming increasingly available. They range from simple vinyl bags for outdoor showers and dishwashing to highly efficient systems using the latest technologies. With increasing documentation that batch heaters can perform well in cold weather and be highly resistant to freezing, the batch heater market is growing across the country.

Figure 2–32.
The Cornell 360 features triple glazing, a selective surface tank treatment, R-16 collector insulation, and well-insulated pipes. It is guaranteed to operate in temperatures down to −10°F. However, Cornell does not recommend use when temperatures fall below 0°F or are below freezing all day for extended periods. According to an independent engineering report, if at night the outside air were 0°F and the water inside the collector had already dropped to 32°F, it would take another 12 hours for the first 3.5 gal. of water to freeze. On clear days, this collector can raise the temperature of the water 70°F to 140°F or more, depending on the time of year. (Courtesy of Cornell Energy, Inc.)

Figure 2–33.
The roof installation of Sunloop's Hot Stuff is accomplished in 10 minutes in new construction; a crane simply places the unit into the recessed roof trusses (patent pending). The plumber has only four copper fittings to connect in addition to the conventional water heater fittings. The roofer then locates flashing around the Hot Stuff as if it were a chimney and the total installation is complete. (Courtesy of Sunloop Corp.)

Figure 2–34.
Servomatic Solar of California is reputed to have sold more passive solar water heating systems than any other manufacturer of solar water heaters—active or passive. (Courtesy of David Bainbridge.)

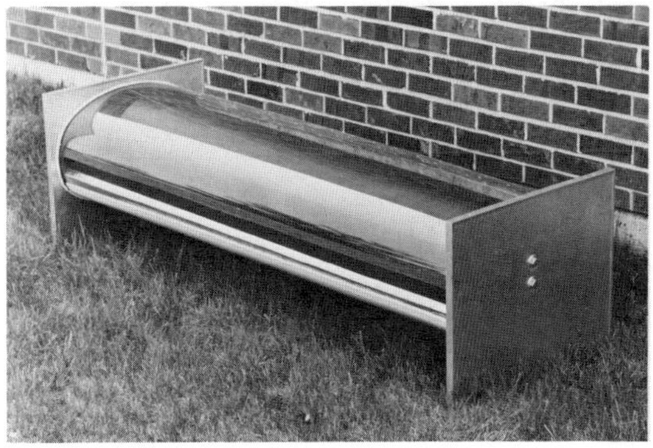

Figure 2–35.
The WTA Solar Water Heater designed by William Tao and Associates uses curved polycarbonate glazing with a UV stabilizer and a 30-gal. galvanized tank.

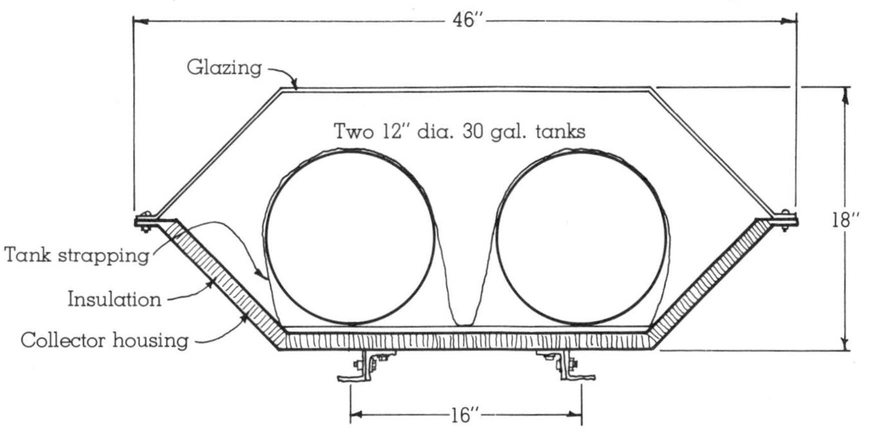

Figure 2–36.
This Integrated Designs collector provides 145°F water on sunny summer days in New England. They are presently developing automatic night insulation for their system. (Courtesy of Integrated Designs.)

Commercial Batch Heaters

Manufacturer	*Product*	*Tanks per unit*	*Tank material*	*Gallons water per unit (liters)*
Basic Designs, Inc. Muir Beach Star Rt. Box 479 Sausalito, CA 94965 (415) 388-5141	Sun Shower	1	vinyl bag	2.5 *(9.5)*
CCMP Solar 1001 C Praire Dr. Austin, TX 78758 (512) 836-2066	SK81	1	n.a., selective surface	30 *(114)*
E.P.I. 1424 West 259th St. Harbor City, CA 90710 (213) 539-8590	Sun Wizard	1	glass-lined water heater	50 is std, other avail. *(189)*
Integrated Designs RFD #2 110 Worcester Rd. Sterling Junction, MA 01565 (617) 422-7312		2	glass-lined steel, selective surface	60 *(227)*
SAV Solar Systems 550 W. Patrice Place Suite A Gardena, CA 90248 (213) 327-7210	SAV, HD 20	1	steel with phenolic resin coating, selective surface	20 *(76)*
Sekisui Chemical Co.	Sekisui Model TM 1000	7	5 ft. x 6 in. OD black	55.5 *(210)*
Sharpe Solar 4300 Easton Dr. Suite 4 Bakersfield, CA 93309 (805) 325-4220	Majic Box	2	n.a., selective surface on second tank	60 *(227)*
Sun Energy International P.O. Box 6542 Concord, CA 94524 (415) 676-8717			plastic cylinders	n.a.
Sunrunner Industries, Inc. P.O. Box 51 Ogdensburg, NJ 07439 (201) 827-8656	Suntube	1	vinyl bag	1 1/2 to 15 *(6 to 57)*

Commercial Batch Heaters

Glazing material	*Internal thermosiphon tank*	*Freeze protection*	*Aperture area sq. ft. (sq. m.)*	*Reflectors*	*Approx. cost*
one side clear	no	no	1.5 (*0.1*)	no	$15
n.a. double	no	no	n.a.	n.a.	$600
fiberglass-reinforced polyester, Tedlar coated honey comb	no	built-in heater	55 (*5.1*)	yes, white wall	$750
single fiberglass	no	no	28	internal	n.a.
double-glazed butyrate	yes	no	24 (*2.2*)	external semi-parabolic anodized aluminum	dist. only
polycarbonate	no	no	67 (*6.2*)	internal	$400
n.a. double	no	no	n.a.	n.a.	n.a.
n.a.		n.a.	n.a.	n.a.	n.a.
clear vinyl	no	no	2 to 16	no	$9.50 to $16

Commercial Batch Heaters

Manufacturer	*Product*	*Tanks per unit*	*Tank material*	*Gallons water per unit (liters)*
Solar American Co. P.O. Box 2088 Newport News, VA 23602 (804) 874-0836	Summersun	1	galvanized steel, hot dipped with selective coating	30 (*114*) 40 (*151*)
Tao-Starr Solar Corp. 1515 Fairview Ave. St. Louis, MO 63132 (314) 423-7898	WTA Solar Water Heater	1	galvanized steel, hot dipped	30 (*114*)
Solar Tube, Inc. Box 12513 Austin, TX 78711 (512) 928-4787	Solar Tube #1	1	glass-lined steel with selective surface	30 (*114*)
Solar Tube, Inc. Box 12513 Austin, TX 78711 (512) 928-4787	Solar Tube #2	1	glass-lined steel with selective surface	30 (*114*)
Sunloop Corp. 5838 Robertson Ave. Carmichael, CA 95608 (916) 486-1801	Hot Stuff™	2	glass-lined steel with selective surface	60 (*227*)
Cornell Energy, Inc. 245 S. Plummer Suite 29 Tucson, AZ 85719 (602) 882-4060	Cornell 360	1	glass-lined steel with selective surface	32 (*121*)
Servomatic Solar 622 Contra Costa Suite 270 Pleasant Hill,CA 94523 (415) 687-0600	Sunflow	1	stainless steel	18.9 (*71.5*)
Energy Engineering 1704 Stanford Dr., NE Albuquerque, NM 97106 (505) 266-9591	Sun Trak	1	glass-lined steel	66 (*250*)

Commercial Batch Heaters

Glazing material	*Internal thermosiphon tank*	*Freeze protection*	*Aperture area sq. ft. (sq. m.)*	*Reflectors*	*Approx. cost*
fiberglass-reinforced polyester (Kalwall)	no	no	10 (*0.9*)	internal	30:$650 40:$675
polycarbonate with UV stabilizer	no	no	16 (*1.5*)	internal	$350-$450
single, fiberglass	no	no	14 (*1.3*)	external	$400
single, fiberglass	no	yes, patented phase change	14 (*1.3*)	external	$1,750
double, heat mirror	no	no	25 (2.3)	no	$2,099
glass with two layers of filon	no	no	18 (*1.7*)	internal	$895
double lexan	no	no	36 (3.3)	external	$3,740
none	heat pipe	heat pipe	44 (4)	yes	$2,000

Source: David Bainbridge, *The Integral Passive Solar Water Heater Book*, The Passive Solar Institute, 1981.

3 BATCH HEATER DESIGN

COMPONENTS

For the most cost-effective batch heater, you must carefully balance the three major aspects of the design: water storage capacity, solar collection, and heat retention. These correspond to the solar water tank, glazing and reflector, and collector enclosure.

Solar Water Tank

The solar water tank (or tanks) is the heart of the batch system. For maximum performance the batch of water must be sized according to hot water use and the amount of glazing. Select the tank(s) based on size, cost, and durability. Because new tanks are one of the most expensive components of the system, used tanks can be considered for a low-cost batch.

To provide 100% solar-heated water during warm weather, the batch should store a one to one and one-half day supply of hot water. Having water in the solar tank the day before it is needed provides solar-heated water for the next morning and cloudy days. But since the benefit from the second day's supply is not always needed, it may be more cost-effective to have only enough water storage for a one-day supply. This is especially true when the second day's water storage means adding another water tank.

Note: When two tanks are used in the batch, plumb them in series with the top of one tank feeding the bottom of the other tank.

When only one day's water supply is stored, strict water conservation can be implemented on cloudy days. Or, a heating element can be installed in the top of the batch tank as an occasional backup heat source.

Batches that store less than a day's hot water demand can be used as preheaters. When used in conjunction with an auxiliary water heater you get almost unlimited hot water. These low-volume batches can be extremely efficient because all of the solar-heated water is used as fast as it can be heated. On the other hand, because the solar-heated water must pass through the auxiliary water tank and additional pipe there is extra heat loss.

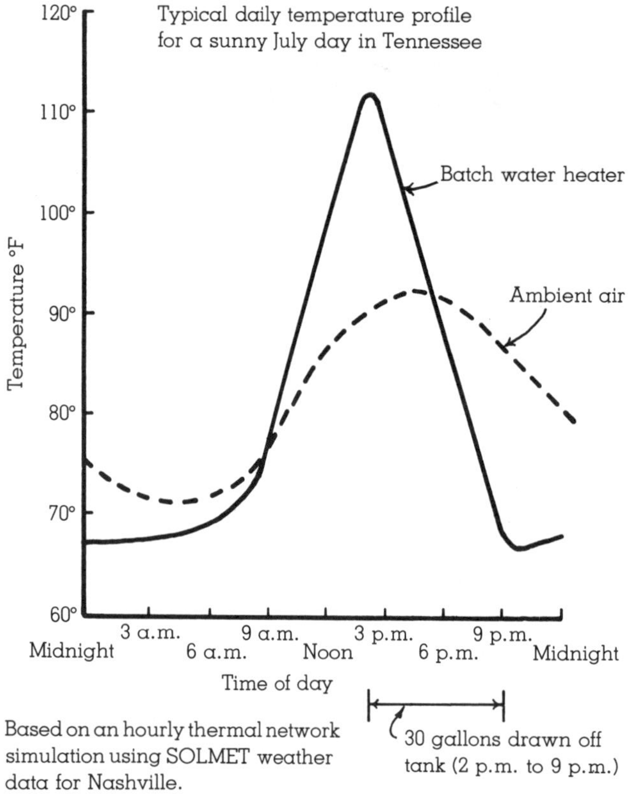

Figure 3–1.
Performance of a 30-gal., single-glazed batch heater. This batch's performance is shown for a sunny summer day when 30 gal. of solar-heated water are used between 2 P.M. and 9 P.M. After 4:30 P.M. the temperature of the water in the batch is lower than the outdoor air temperature; therefore, the water in the batch is heated by the air during the night. (Source: Jim Gruber and Dick Howe, "A Low-Cost Batch Heater," *Solar Age* magazine, May 1982.)

Note: When all of the solar-heated water is used each afternoon, leaving only cold water in the collector at night, elaborate or expensive heat retention is usually not necessary. (The exception is when the batch must be protected from freezing.) On warm nights the cold water in the collector can actually gain heat.

Typical solar water tank storage for a family of three ranges from 30 to 90 gal. Depending on the implementation of conservation devices and personal use of hot water, a person may use anywhere from a staggering 34 to a Spartan 3 gal. of hot water each day. Because of recent high energy costs and resulting awareness of conservation, average daily hot water use is now approximately 20 gal. A few years ago the average was about 30 gal. per day. And, with ever-increasing conventional energy costs making conservation even more cost-effective, many people are in the 10 gal. per day range.

To ensure rapid heating of the water, the solar tank(s) should be long and narrow; 12″ to 16″ in diameter is a good size. This diameter provides a good ratio between the amount of surface area exposed to the sun and the volume of water

Figure 3–2.
This batch design by the Total Environmental Action Foundation in Keene, New Hampshire, features a 30-gal. tank with a 34″ x 76″ glass cover.

being heated. Larger diameter tanks have smaller surface-area-to-water-volume ratios and therefore they heat and cool more slowly. A 12″ diameter tank holds approximately 6 1/2 gal. for each linear foot of tank, while a 24″ diameter tank has twice the surface area of the 12″ diameter tank but holds four times as much water per linear foot.

Water tanks are available made from glass - or stone-lined steel, galvanized steel, and stainless steel. Glass-lined tanks are the most common. With a "sacrificial" anode rod, standard with most glass-lined tanks, they can be expected to last about ten years. Replacing the anode rod periodically will increase the life of the tank (when the glass lining is chipped or cracked, the anode rod corrodes instead of the tank). Stone-lined tanks are not generally used for batch heaters because these tanks cannot be placed at an angle or horizontally, and they are expensive. Also, the stone lining delays the transfer of heat to the water. Stone-lined tanks last about 20 years. Galvanized steel tanks are generally available as range boilers and water storage tanks. Because hot water is more corrosive than cold water, only extra-heavy galvanized steel tanks should be used. These extra-heavy galvanized tanks are not readily available and usually cost more than glass-lined water heaters. Depending on the temperature of the water in the tank, extra-heavy galvanized tanks can last from 10 to 20 years. Stainless steel tanks have the greatest life expectancy of the various tank materials. But their high cost must be weighed against this longevity. Stainless steel tanks suitable for batch heaters have recently become available.

Unfortunately it is usually difficult to purchase a new water tank without the outer jacket, insulation, and heating elements. Removing and discarding these items is certainly a waste of time and materials. But because water heaters are produced in such large quantities, these units can usually be purchased for the same price as a bare tank. Check for sales at local department stores. A 40- to 50-gal. glass-lined water heater with the least insulation and fewest heating elements costs about $150.

Used water tanks are only recommended for the owner-built system. Make certain that the system is designed so that the tank is easily replaceable. Used water heaters are often discarded when their heating elements malfunction, but the tank may still have a few good years of use in it. These tanks can usually be purchased for a few dollars, or sometimes obtained for free.

To check the tank on site, bring a large wrench, hammer, pry bar, penetrating oil, and a flashlight. First, remove the outer jacket and insulation and look closely at the welded seams, which are generally the first place leaks occur. Next, remove the fittings. Fittings that you cannot remove may be left in place if you have ample openings for your plumbing connections. Look inside the tank, shining the flashlight into the tank from another opening. If the glass lining is still intact, the tank is probably worth taking home for cleaning and a pressure test.

Rinse out the tank thoroughly with water and detergent. Pressure test the tank with at least 70 psi of water pressure. If the tank holds the pressure for 24 hours, you have a good tank. Minor leaks can sometimes be repaired with epoxy. Last, check and replace the anode rod if needed.

Glazing

The glazing area, number of layers, and material must be selected. Performance, initial cost, durability, and ease of construction should be considered.

For good solar performance provide between .45 and .65 sq. ft. of glazing for each gallon of water. That is, each square foot of glazing can heat 1 1/2 to 2 1/4 gal. of water. For example, a 40-gal. batch should have between 18 and 26 sq. ft. of glazing area. If the glazing is completely wrapped around the tank, reflectors must be used to shine light through the non-south-oriented glazing. The larger the glazing area for a given volume of water, the faster the water will be heated.

Note: When it is possible to enlarge the glazing or reflector area without significantly increasing the total cost or amount of work, always use the larger area. A large solar collection area can overcome many shortcomings of siting, construction, or design. Also, batches that store only a one-day supply of hot water should use no more than 1.8 gal. per sq. ft. of glazing to ensure rapid heating of the water.

When the batch water supply is large enough to provide hot water at night and into the next day, heat retention is crucial. The glazing is generally the weakest area of heat retention in the batch collector. Double or triple glazing plus additional heat retention schemes such as night insulation and selective surface are recommended (more on performance later).

Each layer of glazing can be selected separately. Some glazing materials are especially well suited to certain batch designs and applications. Use the following table and discussion on recommended glazing systems to assist in the selection.

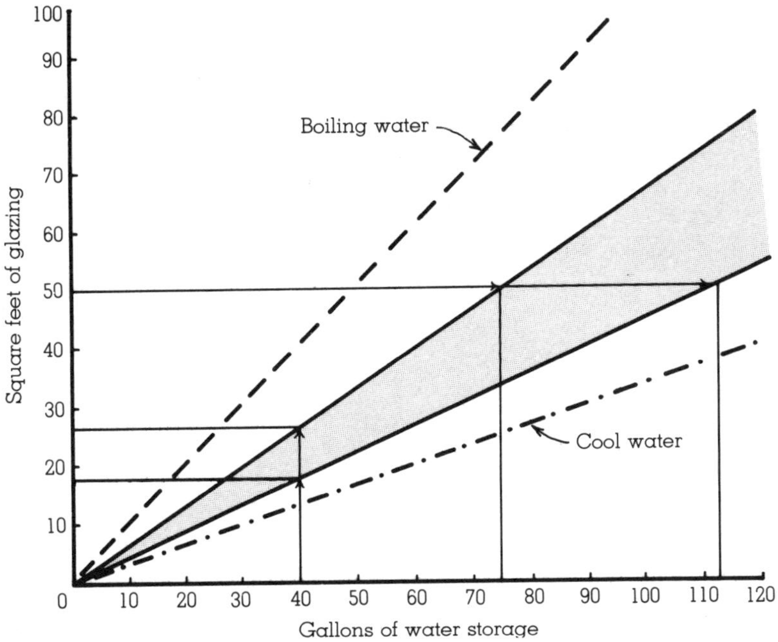

Figure 3–3.
Recommended Ratio of Water Storage to Square Footage of Glazing. For good solar performance, .45 to .65 sq. ft. of glazing should be provided for each gallon of water. For example a batch heater with a 40-gal. tank should use between 18 and 27 sq. ft. of glazing. This ratio can be turned around: with 50 sq. ft. of glazing, for example, between 75 and 113 gal. of water storage should be used. *The area of glazing is defined as the effective area of solar collection for the entire day including effective reflector area.

Figure 3–4.
This batch heater, located in the attic of the solar greenhouse, consists of two 26-gal. tanks and two 46″ x 76″ triple-glazed panels (low-iron glass outer glazing and two layers of Teflon inner glazing). Because there are only 1.1 gal. of water per square foot of glazing area, the water in this batch boils! (Designed by Nature's Way Energy Systems.)

Type of material	*Brand name and manufacturer*	*Thickness*	*Weight/area (lb./sq. ft.)*	*Solar transmittance (%)*
Glass	standard float glass	3/16′′	2.4	84
	low-iron: Sunadex by AFG Industries, Box 929, Kingsport, TN 37662	3/16′′	2.4	91
Acrylic	Plexiglas by Rohm & Haas Co., Independence Mall W., Philadelphia, PA 19105	1/8′′	.7	93
	Exolite (double-glazed unit) Cyro Industries, 897 Route 46, Clifton, NJ 07015	5/8′′	1.0	86 (for double-glazed unit)
Polycarbonate	Tuffak Twin-Wall (double-glazed unit) by Rohm & Haas Co., Independence Mall W., Philadelphia, PA 19105	.22′′	.25	79 (for double-glazed unit)
	Lexan by General Electric	1/8′′	.75	87
Fiberglass Reinforced Plastic	Kalwall Sun-lite by Solar Components, Box 237, Manchester NH 03105	.025′′ .04′′ .06′′	.25	85
	Lascolite	.04′′		81
	Filon by Vistron Corp.	.04′′		82
Films	Tedlar by DuPont, CSC Chestnut Run, Wilmington, DE 19898	4 mil	.03	92
	Teflon by DuPont	1 mil	.02	96

Thermal expansion	*Max. temp. (°F)*	*Estimated life (yrs.)*	*Trans-parent*	*Trans-lucent*	*Cost ($/sq. ft.)*	*Notes*
0.47	400	very long	x		$1.50	Excellent esthetic quality; large double and triple-glazed units are heavy, but otherwise easy to install.
0.47	400	very long	x	x	$2.00 - $4.00	
4.0	180	20	x		$2.00	High transmittance; lightweight; does not yellow. Scratches easily; slight embrittlement with age; large thermal expansion (gaskets are recommended for sealing edges). Exolite is available in 47 1/2″ x 8′, 10′, 12′, (and longer) double-glazed units. Cyro Industries also sells aluminum glazing caps with gaskets to seal the edges.
4.0	180	20		x	$3.00 - $4.00 (double-glazed	
3.3	190	5		x	$2.50 (double-glazed)	High-impact strength. Large thermal expansion; some yellowing with age.
3.3-4.0	270	10	x	x	$4.00	
1.36	225	15		x	$1.00	Available in 4′- and 5′-wide rolls, and factory-made, double-glazed panels. Large thermal expansion makes material from rolls difficult to apply tightly.
1.36	160	15		x	$ .60	
1.36	160	15		x	$ .85	
2.8	150	5		x	$ .65	High transmittance; low cost. Recommended for inner glazings only.
5.9	300	25	x		$ .70	

Recommended Glazing Systems

Glass. Tempered glass is excellent as exterior glazing. The tempering is necessary for added strength. Glass is very long-lasting, durable (except when struck by flying objects), and gives a high-quality appearance. Low-iron glass, when available and affordable, is recommended for its additional (+10%) solar transmittance. If the cost of low-iron glass is much more than 10% greater than regular float glass, then the additional expense is probably not cost-effective. You can tell the difference between low-iron and standard glass by looking at the edge of the glass: standard glass has a faint green tint, low-iron glass is clear. An additional advantage of low-iron glass is that it is often textured. The textured glass is esthetically pleasing and does not allow you to clearly see the tank and inner plumbing.

Glass can also be used as inner glazing. The disadvantages of multiple layers of glass are higher cost and less solar energy transmission than with some plastic glazing materials. Double-glazed glass units are readily available in standard sizes, including 34'' x 76'' and 46'' x 76''. These units are used as patio glass door replacement panels and cost about $60 to $90 per unit. (Glass prices can vary greatly, so shop around. Check at your local lumberyard.)

Using these insulated patio glass door replacement panels for collector glazing can sometimes present problems. Since the collector glazing may be sloped, and also exposed to higher temperatures than a patio door would be, the seal between the two sheets of glass can break. Breakage of the glazing seal can cause fogging between the glass and, in very rare cases, cause the glass to break. The use of silicone or polybutyl seals, which can sometimes be specially ordered, minimizes these seal problems. (Do not use insulated glass units with a welded glass seal. The difference in expansion between the inner and outer glazings will crack this type of seal.)

Acrylic. Acrylic is good for both outer and inner glazings. It is strong, lightweight, and has excellent solar transmittance. Thin sheets of acrylic can be bent for curved glazings.

Exolite by Cyro Industries is a 5/8''-thick, double-walled acrylic glazing. It is available in sheets 47 1/2'' wide and lengths of 8' and longer in 2' increments. (Exolite is also available from the same manufacturer in polycarbonate.) Exolite is rigid and can only be bent along a seam after it has been cut on one side and heated on the other side. Similar to most plastic glazings, acrylic has a large thermal expansion and therefore edges of large sheets should be sealed with gasket material.

A disadvantage of acrylic is that it scratches easily, degrading the appearance of the collector over a period of time.

Fiberglass-Reinforced Plastics. Kalwall, Lascolite, and Filon are some of the many fiberglass-reinforced glazing available. These flexible sheets have high impact strength, which makes them good for outer glazings. Only those that can withstand temperatures of 150°F and above should be used for inner glazings. These sheets are easy to install and are especially good for cylindrical glazing enclosures around a solar water tank. One disadvantage is that they deteriorate somewhat over the years.

Polycarbonate. Polycarbonate can be used for both inner and outer glazings. Its high-impact strength makes it suitable for use in situations where other glazings would be destroyed. Its high cost and the fact that it may yellow slightly after several years are its disadvantages.

Films. Teflon is an excellent inner glazing material combining high solar transmittance and moderate cost. It is recommended for multiple layers of inner glazing. All films are lightweight and must be supported on all sides by a rigid frame. Films are not recommended for outer glazings.

Batch Enclosure

All nonglazed areas of the collector should be heavily insulated. R-20 insulation in mild climates and R-25 in cold climates should be the minimum. To protect pipes inside the collector from freezing, use at least R-8 insulation around them and make certain that the insulation will not degrade from exposure to the sun.

The interior surfaces of the collector should be either reflective or black. All materials inside the collector must be able to withstand air temperatures of 150°F to 200°F. Black surfaces within the collector can become much hotter. Check specifications from materials manufacturers to ensure that the exposed surfaces will not degrade or emit gases at these temperatures.

Note: Where the angle of the glazing and interior reflectors is not designed to reflect most of the sun to the tank throughout the heating season, use a black interior surface to convert the sunlight to heat instead of having the light reflected back out of the collector.

LOCATION

The batch collector can be constructed on the roof, attached to the exterior of the south wall, inside a south-facing room or greenhouse, or detached from the house. Each location can have advantages and disadvantages, depending on your exact situation. Consider ease of construction, structural suitability of the house, esthetics, and solar performance when designing your system.

Roof-mounted solar water heaters have the distinct advantage of often being high above most potential shading obstructions, although tall trees and buildings, especially those nearby, can still be a problem. The collector can be built either above the roof or with the glazing flush with the roof and the tank located in the attic. For new construction and for existing houses that have adequately strong roofs oriented toward solar south, the roof-mounted collector is a good choice.

Figure 3–5.
A roof-mounted batch whose tank sits in unused attic space. (Designed by Nature's Way Energy Systems.)

A major disadvantage of the owner-built roof-mounted collector is that most people are not accustomed to working on a roof. Problems such as carrying materials up to the roof (including a batch heater's 100+-lb. tank and manipulating the materials into position once they are on the roof can be dangerous for the weekend builder. Another consideration of the roof-mounted batch is that the structure of the roof must be able to hold the weight of the collector—approximately 600 lb. for a 40-gal. system and 1,100 lb. for an 80-gal. system. Most roofs designed to hold heavy snow loads can support this weight provided that the load is distributed over a large enough area of the roof. Check with your local building inspector to find the design load of your roof. Otherwise, bracing can sometimes be added to support the weight of the collector. Along with these considerations is the problem of potential roof leaks. Experienced solar installers have mastered these details. For the first-time installer, extreme care and a willingness to go back and try again is often the only solution. When working with an existing roof not oriented toward solar south or not properly sloped for good year-round solar gain, the additional work necessary to construct a special mounting stand is another reason to use an alternate location.

For many owner/builders, mounting the collector on the south wall of the house is an excellent choice. Wall-mounted collectors are attractive and can provide good solar performance. Chapter 5 provides complete step-by-step construction instructions for a wall-mounted unit.

The wall-mounted collector is built so that it becomes a permanent part of the house. By using the same siding on the collector that is used on the house, the sleek

Figure 3–6.
Wall-mounted batch. (Constructed by Anastas Pollock.)

lines of the collector often add to the attractiveness of the house. This is a real benefit for those who do not care for the way some roof-mounted collectors look.

Other potential benefits of having the collector mounted directly on the wall include simplified plumbing and freeze protection, and solar space heating. Plumbing lines between the collector and auxiliary water heater can often be routed between floor joists directly into the collector. This can be a real benefit when adding a batch to an existing house. Also, this usually avoids subjecting the plumbing lines to outdoor freezing temperatures. For freeze protection and supplementary solar heating, the wall separating the collector and house can be equipped with thermosiphoning or fan-powered vents. In cool weather these vents allow warm house air into the collector as a means of freeze protection. During cold weather the solar water tank is drained and hot air in the collector is blown into the house for supplementary daytime space heating.

Since most wall-mounted collectors are constructed at ground level (although the collector could be built on an upper-story wall), the dangers associated with working on the roof are eliminated. Also, with the collector at eye level it can be routinely inspected and maintained as needed.

A wall-mounted unit may not work for you if: (1) your south wall is heavily shaded or you do not have a south wall, a possibility with attached housing, or (2) the few feet that the collector would protrude from the south wall would interfere with the walking path around your house.

Building a batch in a separate structure detached from the house can sometimes overcome these two major disadvantages of the wall-mounted collector.

With a moderate amount of outdoor space and the freedom to construct your collector anywhere in your yard, you will probably be able to find an unshaded

Figure 3–7.
Ground-mounted batch supported on concrete posts. (Constructed by Greg Shinberg.)

Figure 3–8.
This horizontal batch tank is located in an upper greenhouse space. The white wall behind the tank reflects extra light onto the tank. The greenhouse glazing is at a 60° angle and the tank has no glazing around it. On sunny days the water temperature in the tank is raised 30°F to 40°F. The warmed water is used to water the plants in the greenhouse and to preheat water for the auxiliary water heater.

area as well as orient the collector close to solar south. Also, the collector can be built so that it does not interfere with other activities near the house.

The disadvantage of the detached collector is that additional time and money may be required for installing the collector's foundation and extra plumbing. The detached collector should be heavily insulated on all sides but south since it is totally exposed to the outdoors. The exception to this is the batch that is completely enclosed with glazing and placed in front of a reflector.

Placing your batch heater indoors is only recommended in cold climates. Indoor batches have the unique advantage of being completely sheltered from the outdoor environment, but they sacrifice sunlight that must first pass through the house glazing and is shaded by the house structure. When designing an indoor batch, great care must be taken to ensure that it receives enough sunlight throughout the year. Large reflectors are necessary to compensate for the reduced amount of light. However, since indoor collectors are completely protected from the rain, wind, and cold, they can be extremely simple and low in cost, even in frigid climates.

A disadvantage of locating your solar water heater indoors is that it can take valuable space unless you skillfully build it into unused space, such as a window seat or overhead behind a skylight. Also, allowing additional sunlight into your house in summer may cause overheating unless you have excellent ventilation.

Chapter 6 shows how to build a vertical batch that can be used indoors.

COMPARATIVE PERFORMANCE

John Burton of Integral Design, Santa Rosa, California, and Peter Zweig of Farallones Institute, Occidental, California, have studied the performance of batch water heaters at the Farallones Institute's Rural Center. Their tests compared the performance of batches with variations in:

1. glazing—number, type, and treatments
2. tank treatment—flat black paint, selective surface, and transmissive insulation
3. interior collector surfaces—reflective, white, and black

The results of their study provide much of the basic information you need to design your own batch water heater.

The relative thermal performance of two sloped, 30-gal. systems with various designs was compared. Each collector was sloped at a 45° angle, and the collector box had R-19 fiberglass insulation in the walls and R-11 fiberglass insulation in the floor. The effective glazing area of the test collector was 17.5 sq. ft., which provides a ratio of 1.7 gal. of water per sq. ft. of glazing.

The test results shown below are given for a clear day after each collector had been filled and placed in the sun for several days without water being removed or added. Had the collector been in use with cold water frequently being added to the

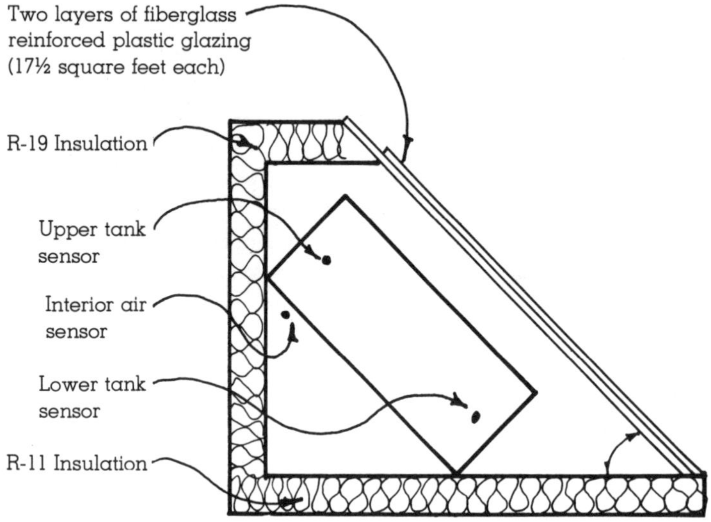

Figure 3–9.
30-gal. batch solar water heater used by John Burton and Peter Zweig in performance tests.

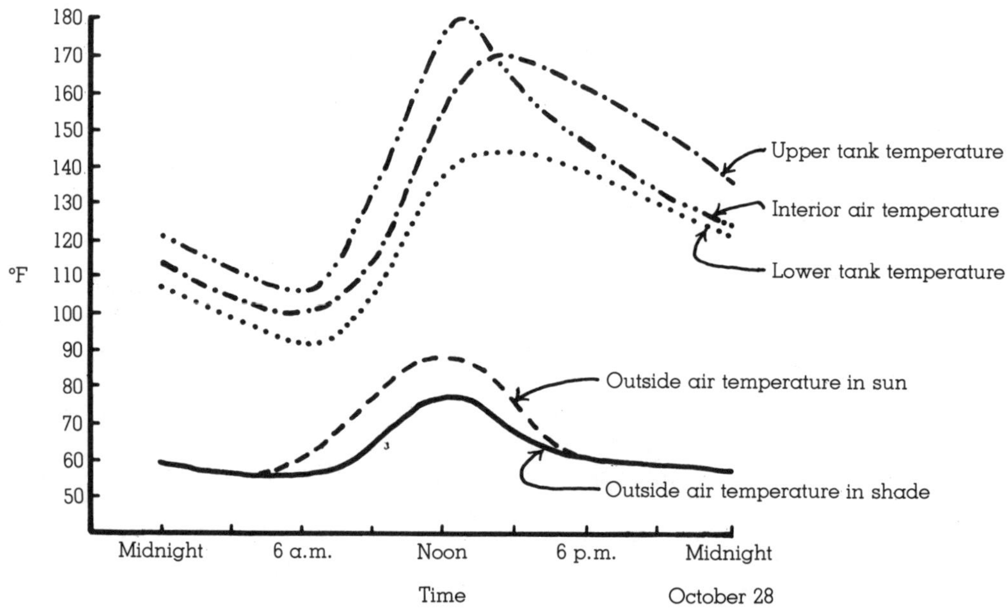

Figure 3–10. Thermal performance of a bath consisting of a flat black tank and two layers of fiberglass-reinforced glazing. (Source: John Burton and Peter Zweig.)

collector, then the temperatures would have been lower than those shown in these tests. The temperatures shown are from the maximum/minimum outside air thermometer and the upper tank water sensors. The outdoor maximum temperatures occur in the afternoon, before the solar tank maximum. The outdoor minimum temperatures occur about sunrise. Figure 3–10 shows typical batch and outdoor temperature fluctuations for an entire day in late October.

Batch Water Heater Comparison Tests

		Low	*High*	*Low*
Test 1 10/5/80–10/6/80		9:00 A.M.	4:00 P.M.	9:00 A.M.
outside air		54°	79°	50°
no glazing	flat black tank	70°	122°	64°
one sheet of fiberglass	flat black tank	102°	140°	95°
Test 2 10/17/18–10/18/80		9:00 A.M.	4:00 P.M.	9:00 A.M.
outside air		46°	72°	45°
one sheet of fiberglass	flat black tank	90°	144°	90°
one sheet of glass	flat black tank	97°	149°	97°
Test 3 10/29/80–10/30/80		8:00 A.M.	3:00 P.M.	8:00 A.M.
outside air		50°	75°	50°
one sheet of fiberglass	flat black tank	93°	141°	91°
two sheets of fiberglass	flat black tank	109°	156°	109°
Test 4 11/6/80–11/7/80		8:00 A.M.	3:00 P.M.	8:00 A.M.
outside air		52°	75°	54°
one sheet of fiberglass with R-12 night insulation	flat black tank	93°	144°	111°
two sheets of fiberglass	flat black tank	95°	145°	109°
Test 5 11/13/80–11/14/80		7:30 A.M.	2:30 P.M.	7:30 A.M.
outside air		39°	68°	43°
one sheet of fiberglass	flat black tank	81°	136°	82°
three sheets of fiberglass	flat black tank	100°	149°	106°
Test 6 12/14/80–12/15/80		8:00 A.M.	2:30 P.M.	8:30 A.M.
outside air		34°	69°	39°
one sheet of fiberglass	flat black tank	75°	126°	75°
one sheet of fiberglass	transmissive insulation	95°	124°	88°

Test 7 1/14/81–1/15/81		8:30 A.M.	3:00 P.M.	8:30 A.M.
outside air		43°	64°	43°
two sheets of fiberglass	transmissive insulation	82°	117°	90°
two sheets of fiberglass	selective surface	93°	129°	100°
Test 8 2/3/81–2/4/81		**8:30 A.M.**	**3:30 P.M.**	**9:30 A.M.**
outside air		43°	64°	43°
two sheets of fiberglass	flat black tank	99°	151°	100°
two sheets of fiberglass	selective surface	115°	156°	117°
Test 9 2/21/81–2/22/81		**8:00 A.M.**	**4:00 P.M.**	**8:30 A.M.**
outside air		46°	73°	46°
two sheets of fiberglass	flat black tank	102°	160°	108°
two sheets of fiberglass	selective surface and transmissive insulation	109°	144°	117°
Test 10 3/5/81–3/6/81		**8:00 A.M.**	**4:00 P.M.**	**8:30 A.M.**
outside air		39°	52°	45°
three sheets of fiberglass	flat black tank	75°	140°	102°
two sheets of fiberglass	selective surface and transmissive insulation	79°	120°	100°
Test 11 3/9/81–3/10/81		**8:00 A.M.**	**4:00 P.M.**	**8:00 A.M.**
outside air		45°	70°	46°
three sheets of fiberglass	flat black tank	100°	153°	111°
two sheets of fiberglass	selective surface	97°	147°	117°
Test 12 5/3/81–5/4/81		**9:00 A.M.**	**4:30 P.M.**	**9:00 A.M.**
outside air		43°	72°	45°
two sheets of fiberglass	flat black tank	102°	142°	102°
two sheets of fiberglass	flat black tank and half transmissive insulation	118°	142°	118°

Note: The results of this study were first published in the *Proceeding of the Sixth National Passive Solar Conference*, Portland, Oregon, 1981. They are reprinted here with the permission of John Burton and Peter Zweig.

Glazing In the first test the use of a single sheet of fiberglass-reinforced glazing increased the tank temperature approximately 30°F over the temperature of the unglazed tank. The second test showed that a single sheet of 3/16″ thick tempered glass produced a 5° to 7°F advantage over a single sheet of fiberglass-reinforced glazing. In test 3 two layers of fiberglass-reinforced glazing produced temperatures approximately 15°F higher than those produced by a single layer of fiberglass-reinforced glazing.

The use of night insulation over glazing can be effective in reducing heat loss, as shown in test 4. Here a single-glazed fiberglass collector with nighttime R-12 rigid insulation showed a slight advantage over a double-glazed fiberglass collector. But the advantage was realized only when the shutter was added at the end of the daily solar gain and removed when the first morning light hit the collector. This is difficult to manage, even for the most enthusiastic solar advocates.

In test 5 a triple layer of fiberglass showed a great reduction in night heat losses over a single layer of fiberglass-reinforced glazing, while equaling the afternoon heat gain allowed by two layers of fiberglass-reinforced glazing in previous tests. The use of multiple glazings instead of manually operated shutters to reduce night heat loss frees you from daily attention to the collector.

Tank Treatments

For years, Horace McCracken has successfully used fiberglass furnace filter material wrapped around the solar tanks to reduce night heat loss. This *transmissive insulation* allows a good percentage of the incident sunlight to strike the tank while providing insulation around the tank. In the sixth test, a flat black tank wrapped with transmissive insulation (1″ thick) showed a slightly lower daytime gain and a strong reduction in nighttime loss, when compared with a collector whose tank was simply painted flat black. Both collectors used single layers of fiberglass-reinforced glazing.

During a sunny day, the single-glazed collector with transmissive insulation had interior air temperatures continuously 20°F above a single-glazed, flat black collector. This hotter interior air (up to 150°F in December) heats the tank during the day and reduces heat loss from the tank in the late afternoon. With the added insulative effect of the transmissive insulation, the tank with single fiberglass-reinforced glazing and transmissive insulation had temperatures higher than those of the single-glazed, flat black tank by around 5 P.M.

A selective surface foil, which is highly absorbtive of sunlight and emits little heat, can be a very effective tank treatment. A recently available nickel chrome selective surface with an adhesive backing was applied to one of the tanks in test 7. The other tank was wrapped in transmissive insulation. The tank with the selective surface maintained an 11°F advantage over the tank with the transmissive insulation as their temperatures rose and fell at the same slow rate during consecutive sunny days.

In test 8 the tank with the selective surface was always warmer than the tank painted flat black. The thermal advantage was greater at night and in the early morning, while its rate of heat gain was somewhat lower during the day. The reduced daytime gain may be due to the insulative qualities of the layer of adhesive. As with the transmissive insulation, the collector with the selective surface had air temperatures that were significantly higher during the day. These high temperatures are due to the selective surface being highly reflective to the thermal energy in the air surrounding the tank. Solar heat that is transmitted through the glazing but not absorbed by the tank builds up inside the collector and reduces heat loss from the tank.

The combination of selective surface and transmissive insulation in test 9 provided the strongest reductions in night heat loss, but also the lowest rate of temperature gain during the day. With upper tank temperatures that never fell below 100°F during five days of varied weather in February, this combination shows great promise.

Collector Interior

There has been some debate regarding the advantages and disadvantages of making the collector interior reflective, black, or white. It turns out that the overall thermal performance between reflective and black interiors is so small that it is hard to detect with the slight inaccuracies of the monitoring equipment. In other tests conducted by John Burton and Peter Zweig white collector interiors yielded top tank temperatures continuously equal to those of a reflective-foil-lined collector box. However, the afternoon air temperatures in the white box were 4°F higher and the bottom tank temperatures 5°F lower than in the foil-lined box. Due to the resulting lower average tank temperature in the white box, the reflective foil seems preferable.

The next comparison was between the reflective-foil-lined collector and the flat black collector. While the top tank temperatures of the black box were always 1°F to 5°F higher than those in the foil box, the lower tank temperatures were 1°F to 4°F lower during the day, and the interior air temperatures were up to 15°F higher, making the overall thermal winner even more difficult to discern. It is a virtual tie for useful energy. Tests with more adverse sky and weather conditions may show a real winner, but it is not likely.

Combinations

As shown in test 10 on page 60, a collector with three fiberglass glazings and a flat black tank was compared with a collector with two fiberglass glazings, selective surface tank, and transmissive insulation. During consecutive sunny days, the triple-glazed, flat black collector had temperatures consistently above the other collector. But after cloudy weather with cold nights the double-glazed collector with selective surface and transmissive insulation had a 13°F advantage. With the transmissive insulation removed from the double-glazed collector in test 11, its solar gain increased to closely match that of the triple-glazed, flat black tank collector. During periods of reduced sunshine the collector with the selective surface stayed warmer.

In test 12 the upper half of a flat black tank was wrapped with transmissive insulation. During the day the upper tank temperature of this double-glazed collector would get up to 16°F hotter than the double-glazed collector without the

Glazing	Tank Treatment	Solar Gain (1—high, 7—low)	Night Loss 1—low, 8—high)
one layer of fiberglass	flat black	1	8
one layer of glass	flat black	1	7
one layer of fiberglass with R-12 night insulation	flat black	1	6
two layers of fiberglass	flat black	1	6
three layers of fiberglass	flat black	1	5
one layer of fiberglass	selective surface	2	6
two layers of fiberglass	selective surface	2	4
three layers of fiberglass	selective surface	2	3
two layers of fiberglass	flat black and half transmissive insulation	3	5
one layer of fiberglass	flat black and transmissive insulation	4	6
two layers of fiberglass	flat black and transmissive insulation	5	4
two layers of fiberglass	selective surface and transmissive insulation	6	2
three layers of fiberglass	selective surface and transmissive insulation	7	1

transmissive insulation. The bottom tank temperatures of both of these collectors stay about equal. This thermosiphoning technique shows much promise.

From these tests, Burton and Zweig rated all of the collector designs in order of effective solar heat gain and night heat loss.

In general, the solar water heater with the highest daytime gain and the lowest night loss will provide the best performance. Creative combinations of passive materials and techniques, such as differing treatments for the top and bottom or front and back of the solar tank, may yield the best performance. Some batch heaters use several tanks to minimize the mixing of incoming cold water with the solar-heated outlet water during water use. These multitank systems could incorporate various combinations of glazing, tank, and collector box strategies. *Maximizing the solar gain should have the highest priority for the cooler first tank, while minimizing the heat loss would be more important for the hotter output tank.*

A primary factor affecting solar water heater efficiency is the size and time of the hot water demand. If most of the hot water use occurs in the afternoon, then the rate of night heat loss is of lesser importance. If, on the other hand, most of the hot water use occurs in the morning, then reduced night heat loss is of primary importance.

Finally, the cost-effectiveness of any solar water heater design must be considered. The real advantage of a batch heater is its simplicity and resulting low cost. Performance-improving materials and their costs are listed below.

Batch Heater Materials and Costs (1982 prices)

Material	*Cost ($/sq. ft.)*	*Cost for a system with two 40-gal. tanks*
fiberglass	$1	$40 per layer
tempered glass	$1 to $2.50	$40 to $100 per layer
R-12 night insulation	$2.50	$100
transmissive insulation	$2	$5 per tank
selective surface	$2.50	$50 per tank

The selective-surface tank treatment provides one of the best overall efficiency improvements for the solar water heater, but it is expensive. Transmissive insulation, on the other hand, is low in cost and, when used strategically, is very cost-effective. In the past, Burton and Zweig have recommended using at least two layers of glazing as cost-effective for a batch heater in most California climates. Based on this research, they have found that double glazing is just a starting point for cost-effective, high-performance batch heaters.

SYSTEM DESIGN WORKSHEET

This worksheet is provided to assist you in designing the best water heating system for your needs. First, circle the appropriate letters on the Analysis section of the worksheet. After you have answered all the questions, match your answers with the letters in the Selection section. Base your final decision on a complete reading of this book and on the tastes and preferences of your household.

Analysis

Site Analysis

1. Orientation
   a. within 20° east or west of solar south
   b. between 30° and 60° east or west of solar south
   c. more than 60° away from solar south

2. Shading
    a. less than 20% of the sun between 9 A.M. and 3 P.M. is shaded from the collector
    b. more than 20% of the midday sun is shaded from the collector

Architectural Analysis

1. South wall
    a. light colored
    b. flat unobstructed area at least 3′ 6″ by 7′ 7″
    c. solar greenhouse or sunroom
2. Roof
    a. flat
    b. sloped toward south
    c. not sloped toward south
    d. very steep
3. Outdoor area
    a. large
    b. limited or no space

Climate

1. warm
2. cool
3. cold

Economics—amount of money you want to invest in solar, including tax rebates and financing charges

1. under $100
2. up to $800
3. $1,000 plus

Selection

Site Analysis

1. Orientation
    a. excellent solar orientation
    b. suitability of site will depend on the slope of the collector, use of reflectors, and climate; see page 67 for guidelines
    c. poor orientation; another side of the building is oriented close to solar south
2. Shading
    a. shading will not significantly affect collector performance
    b. collector performance will be noticeably diminished

Architectural Analysis

1. South wall
   a. wall can be used as a reflector for a vertical tank collector
   b. collector can be mounted on the south wall using plans in chapter 5
   c. collector can be constructed indoors (recommended only for cold climates) or built into the room
2. Roof
   a. ideal location for collector
   b. good location for collector—consider glazing-wrapped solar tank with exterior reflector
   c. additional work and expense needed for construction of collector mounting system
   d. not a good collector location except for experienced builders
3. Outdoor area
   a. appropriate for a collector detached from the house
   b. use south wall or roof-mounted collector

Climate

1. double-glazing can be used effectively for year-round solar water heating
2. year-round solar water heating requires triple-glazing and selective surface tank covering or night insulation
3. consider combined solar water and space heating system, or indoor batch

Economics—amount of money to be invested

1. use recycled materials
2. build a deluxe collector with top-quality materials
3. buy a commercial or contractor-built system

SITE ANALYSIS

Your site is one of the most important factors determining both the best placement of your solar collector and the amount of heat your system will produce. There are two criteria for determining the solar potential of the collector location: *the orientation of the collector toward solar south,* and *the amount of shading on the collector throughout the year.* The information you collect from studying your potential collector site should be added to the Analysis Section of the System Selection Worksheet on page 64.

ORIENTATION

At midday, go outside and stand at the south side of your house. To find *solar south* you must first find out at what times the sun rises and sets on that day. Newspapers often list these times. Exactly halfway between sunrise and sunset, the sun's east to west position is directly *solar south*. Solar south is also called *true south*. The shadow of a vertical stick at midday points directly away from solar south.

If you live near a time change zone line, actual sunrise and sunset in your area may vary as much as one half hour from that stated for your time zone. To check the direction of solar south, the shadow of a vertical stick is shortest when the sun is directly solar south.

You can also find solar south by using a compass, but you must compensate for the fact that the direction of solar south can vary from magnetic south. Use the Magnetic Variation Map (Figure 4–1) to find the variation between magnetic and solar south for your area.

Note: Always take compass readings from several locations because if there are any large metal objects or heavy electrical lines above or below the ground near where you are standing, these objects can significantly affect your compass reading. Even a metal belt buckle close to the compass can affect the reading; and you don't want to build the collector facing in the wrong direction!

First, find your location on the Magnetic Variation Map. Next, note the direction and number of degrees of variation between magnetic and solar south. If

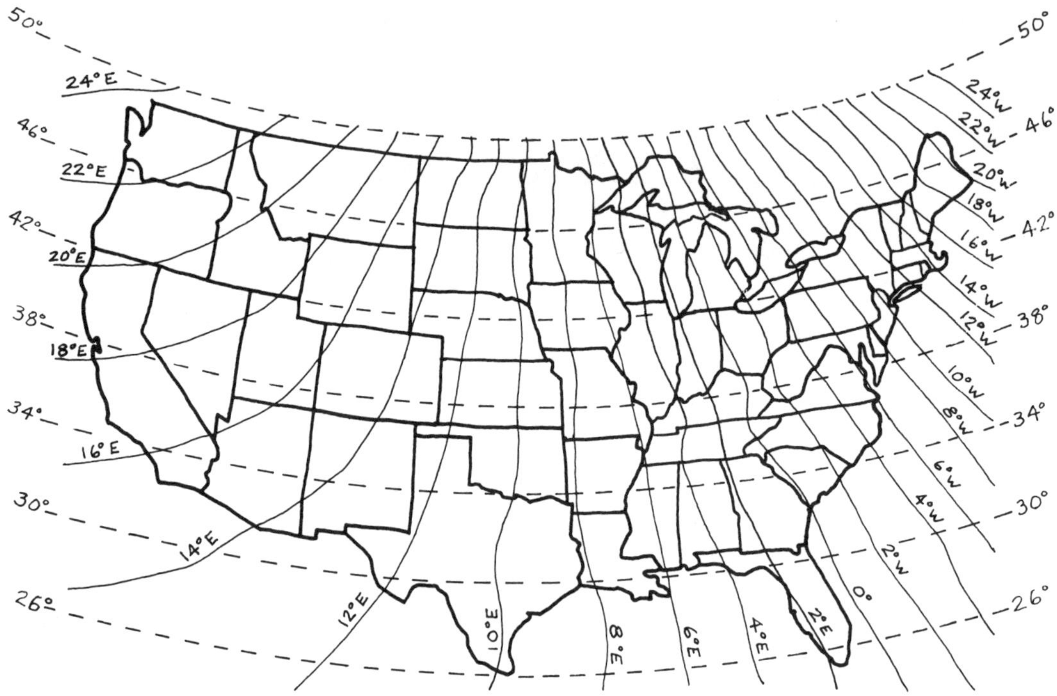

Figure 4–1. *Magnetic Variation Map.* This map of the United States indicates magnetic variation away from solar south. (Source: Isogonic Chart of the Unites States, U.S. Department of Commerce, Coast and Geodetic Survey, 1965.)

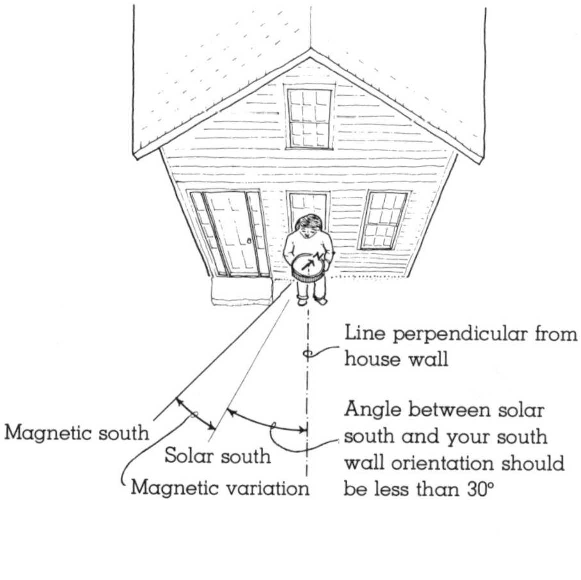

Figure 4–2. Finding the orientation of the south wall or roof.

you live near the line of zero variation, running from Lake Michigan to Georgia, magnetic south and solar south are in the same direction. If you live *east* of this line, magnetic south is to the *left* (east) of solar south; and if you live to the *west* of this line, magnetic south is to the *right* (west) of solar south.

Once you have located solar south, stand with your back in line with the proposed collector. For a south-wall-mounted collector, stand against the south wall. Point one arm straight ahead and the other arm toward solar south. Both arms should be at the same height. The size of the angle between your arms is the number of degrees your south wall is oriented away from solar south.

The orientation of the collector toward solar south must be combined with the slope of the collector to determine the percentage of maximum sunshine the collector will receive in a given location. (See the sample tables for Boston, Miami, Los Angeles, and Seattle.)

Percent of annual maximum performance for Boston (Blue Hill), Massachusetts (42° North Latitude)

	Collector Orientation			
Collector Slope	0°	*30°*	*60°*	*90°*
Latitude-15	99	97	91	83
Latitude	100	98	91	80
Latitude+15	95	93	87	74
Vertical	61	60	55	46

Percent of annual maximum performance for Miami, Florida (26° North Latitude)

	Collector Orientation						
Collector Slope	*0°*	*30°*	*60°*	*90°*	*120°*	*150°*	*180°*
Latitude-15	98	98	96	94	92	90	89
Latitude	100	99	97	93	86	80	78
Latitude+15	97	96	94	88	79	68	62
Vertical	53	53	51	45	29	—	—

Percent of annual maximum performance for Los Angeles, California (34° North Latitude)

	Collector Orientation			
Collector Slope	0°	*30°*	*60°*	*90°*
Latitude-15	97	96	92	87
Latitude	100	99	94	87
Latitude+15	97	96	92	83
Vertical	58	58	55	47

Percent of annual maximum performance for Seattle, Washington (48° North Latitude)

	Collector Orientation			
Collector Slope	*0°*	*30°*	*60°*	*90°*
Latitude-15	100	98	93	86
Latitude	100	98	92	83
Latitude+15	93	92	86	76
Vertical	60	61	59	52

Source: Fuller, W. "Collector Location: No Taboos on East or West," *Solar Age*, December 1980. Reprinted with permission.

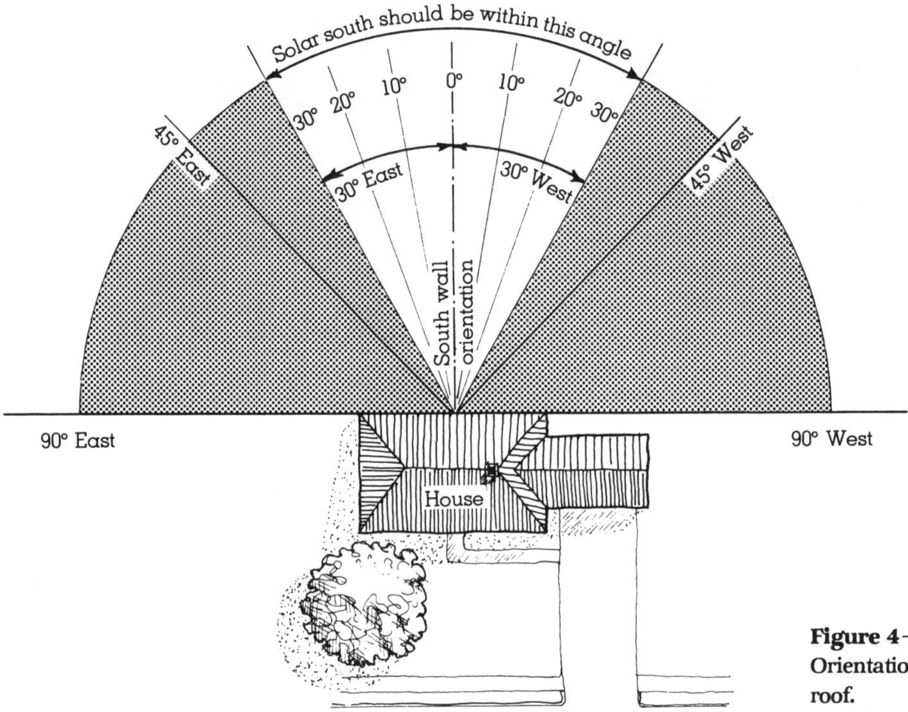

Figure 4–3.
Orientation of the south wall or roof.

For maximum year-round performance, the collector should be oriented toward solar south and be sloped at an angle equal to your latitude. Fortunately, if the collector is oriented within 30° east or west of solar south, collector slopes at angles plus or minus 15° from your latitude angle still receive over 90% of the maximum amount of solar energy.

Therefore, most houses have a wall or roof that is suitable for solar collection. With freestanding solar heating systems, just aim the collector toward solar south. And, if the reflector is separate from the collector, then the reflector should be oriented toward solar south with the collector directly in front of it.

Collectors that are sloped at angles *less* than the latitude angle (more horizontally) receive more sunlight in summer and less in winter. Collectors sloped at angles *greater* than the latitude angle (more vertically) receive more sunlight in winter and less in summer. In general, collectors that are oriented away from solar south benefit from being sloped more horizontally rather than more vertically than the latitude angle.

Since reflectors can significantly add to the amount of solar energy striking the collector, reflectors are an excellent way to achieve good performance with less than optimum orientation and slope. This allows the flexibility to construct a collector that is more easily built and still performs well. In the freestanding vertical passive water heater design (Chapter 6), reflectors can more than offset the loss of efficiency due to the vertical slope of the collector by increasing the amount of solar energy striking the collector by a factor of two or more times.

Slightly easterly or westerly collector orientations can have unique advantages or disadvantages, depending on your exact site. An easterly orientation usually has the advantage of supplying warm water earlier in the morning, while a westerly exposure gives you a little extra heat just before the cold night arrives. Of course, local weather conditions such as frequent morning fogs or afternoon clouds can negate these advantages.

Shading of either the morning or afternoon sun is another consideration in determining a collector's optimum orientation. When the morning sun is shaded, a slightly western orientation is preferable, while shading of the afternoon sun makes a slightly eastern exposure ideal. When building a new house or an addition, orient it to within 30° of solar south so that you not only achieve maximum solar water heating performance, but are also correctly oriented for solar heating and cooling of your house, and for photovoltaics—solar electricity!

SHADING

Many an otherwise fine site has been avoided because it is shaded. Shading of the sun's rays can be a monumental problem for both solar retrofits and new construction. You need to determine which areas of the south wall, roof, or ground are *not shaded from the midday sun throughout the year*.

While standing where the collector would be, imagine the path of the sun as it rises in the east, reaches its highest position at midday, and sets in the west. Actually in winter the sun rises in the southeast and sets in the southwest; and in summer the sun rises in the northeast and sets in the northwest. Only on March 21 and September 21 does the sun rise, then set, exactly east and west. Of course, the sun only appears to rise and set; it is the earth that is rotating toward (rise), and away (set) from the sun.

Shading of the morning or late afternoon sun is not as crucial as shading of the strong midday sun. Winter shading from decidous trees is less than winter shading from evergreens or from solid obstructions, such as the house next door. If you are evaluating your site in summer, spring, or fall, remember that shadows cast by the low winter sun will be *much longer* than those cast during the other seasons. If your potential site has shading of no more than 20% of the sun between 9:00 AM and 3:00 PM and your collector orientation is within 30° of solar south, you have a good solar site!

Since the low winter sun casts the longest shadows, it is most likely to be blocked from reaching your collector. If the shadow is cast by a deciduous tree, then at least some sunlight (about 25% to 50%) will penetrate through the bare tree. Shading of the winter sun is usually not critical for good year-round performance from your solar water heater. (Shading of the winter sun's rays is crucial, however,

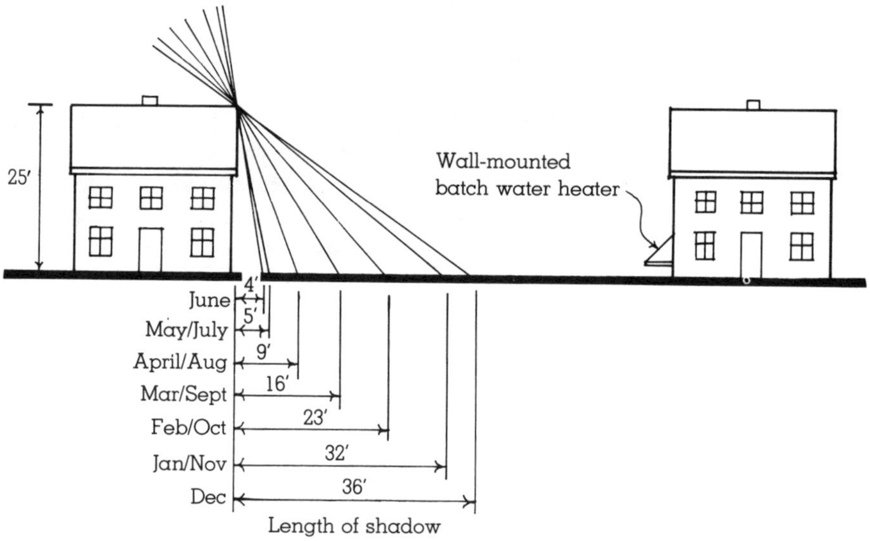

32° North Latitude

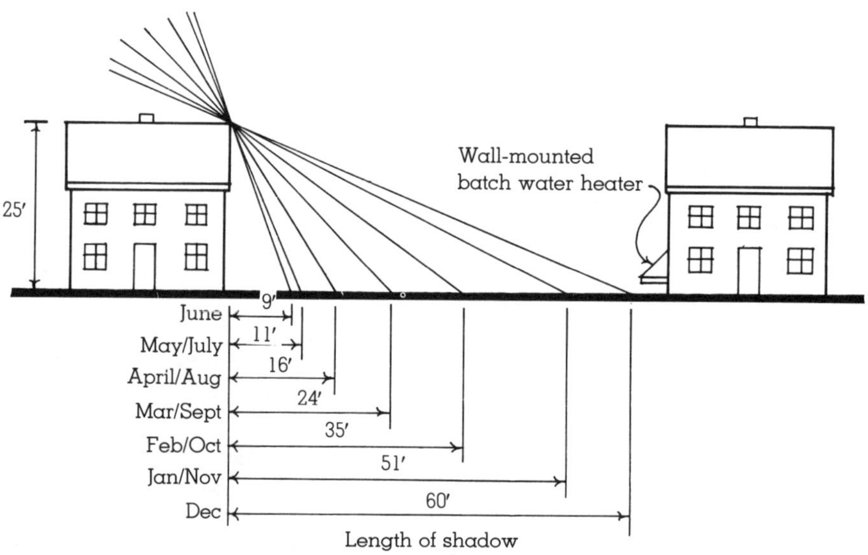

44° North Latitude

Figure 4–4. Length of the shadow cast at solar noon for the twenty-first day of each month by an obstruction 25′ high, for latitudes 32° north and 44° north. A 50′ high obstruction casts a shadow twice as long.

Figure 4–5.
The Solar Card is an easy-to-use, low-cost solar evaluation tool. It can also be used to determine how much sun strikes a potential garden spot or where to plant to let summer shade and winter sun fall on the house.

for solar heating your house.) This is because the amount of sunshine during winter is only a small portion of the year-round potential sunshine.

If you are uncertain about the amount of shading on your collector or want a more exact sighting of the sun's path, you can construct a *solar siting mask.* This device lets you see the winter, fall, and spring sun paths. Because the summer sun is so high in the sky, almost overhead, this siting mask cannot project the sun's path in summer. You will have to approximate the summer sun's path in order to determine any summer shading. Instructions for constructing a solar siting mask are given at the end of this section.

Commercially constructed siting devices are available from:

Solar Card, Design Works, Inc., P.O. Box 489, North Amherst, MA. 01059 ($12.95 postpaid);

Solar Site Selector, 105 Rockwood Drive, Grass Valley, CA 95945 ($89.50 + $4 shipping);

Solar Pathways, Inc., 3710 Highway 82, Glenwood Springs, CO 81601 ($144.00).

The amount of shading varies as you move around your site. Shading from nearby obstructions such as bushes, a chimney, a roof overhang, or a protruding part of your house changes considerably if you move several feet sideways, or up or down.

In new construction, shading from your neighbor's house or trees can cause you to position your building to the north side of your site. When there is no space left to go north, raising the collector a few feet is sometimes as good as moving the building north by twice the distance.

In retrofitting, where you usually have a lot less flexibility, pruning and tree cutting have often been solutions. But, please do not cut down a 100-year-old tree to decrease your shading by a few percentage points.

There is always some frustration after you have decided to go solar— and are faced with the sometimes unwilling site. When building on a compromised solar site, you can sometimes overcome obstacles by using reflectors, or you may simply have to rely on other renewable energy sources and strong water conservation measures.

Sites without any shading on the collector area and optimum orientation are rare. Since there are few perfect solar sites, you must work to integrate the collector with the best qualities of your site.

SOLAR SITING MASK

A solar siting mask is an excellent tool for performing a thorough shading analysis and is especially helpful for evaluating questionable solar sites. The mask can be constructed in several hours with material costs of about $5. When you are finished with your site analysis, the mask is a great gift for a friend.

Construction

Materials

1. 14 1/2'' long, 1'' x 8'' (actual measurement 3/4'' x 7 1/4'') pine board
2. 12'' long, 1/2'' diameter wood dowel
3. 23'' x 16'', .005-mm (or thicker) clear acetate or 1/16'' plexiglass sheet
4. twelve 1/2'' pan head wood screws with 1/2'' washers
5. 23'' x 2'' piece of cardboard

Tools

1. drill and 1/16'', 1/4'', and 1/2'' drill bits
2. sabre saw or keyhole saw
3. black marking pen
4. screwdriver
5. ruler
6. string (for drawing 14 1/2'' diameter circle)
7. glue
8. Six 8 1/2'' x 11'' sheets of 1/8'' graph paper
9. Sun Path Chart (page 77) and Magnetic Variation Map (page 68)

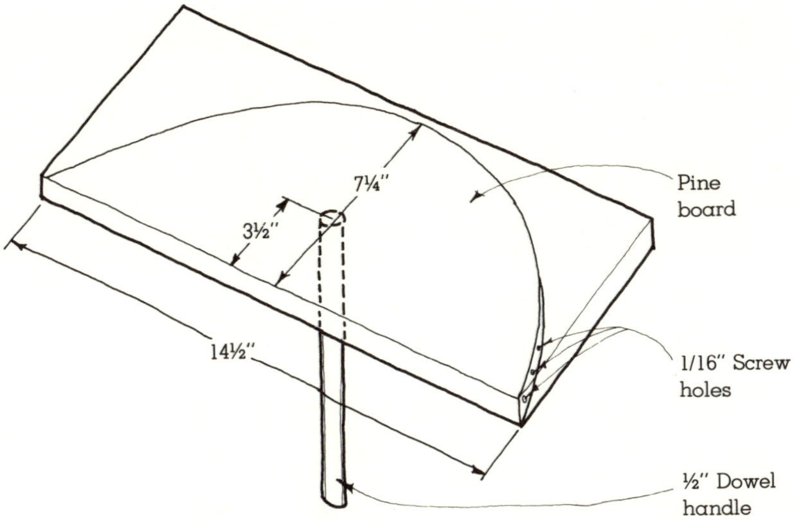

Figure 4–6.
Solar siting mask base and handle.

Constructing the Base and Handle

1. Using a pencil tied to a string, draw a 14 1/2′′ diameter semicircle on the 1′′ x 8′′ x 14 1/2′′ board.
2. Mark the point 3 1/2′′ from the edge and equidistant from the sides.
3. Cut the board to form a semicircle.
4. Drill a 1/2′′ hole at the center point of the board, and glue the end of the 1/2′′ dowel into this hole to make the handle.
5. Drill 1/16′′ holes every 2′′ along the curved edge of the semicircle as shown.

Drawing the Sun Paths

1. Join six 8 1/2′′ x 11′′ sheets of graph paper to make one 22′′ x 24′′ sheet. The 24′′ side should be held horizontally.
2. Find the Sun Path Chart for the latitude nearest yours. You can determine your latitude from the Magnetic Variation Map.
3. Using a ruler, draw the Solar Altitude/Solar Azimuth Graph according to the dimensions in figure 4–7. Solar altitude is the height of the sun above the horizon. Solar azimuth is the east-to-west position of the sun. Note that the solar altitude spacing varies for each 10° interval. This is because when the sun's circular and overhead paths are projected and drawn onto a flat surface, such as the shading mask, the paths are distorted.
4. Draw the Sun Path Chart for your latitude onto the 22′′ x 24′′ graph paper. The easiest way to do this is to mark a dot on the graph paper on each solar azimuth

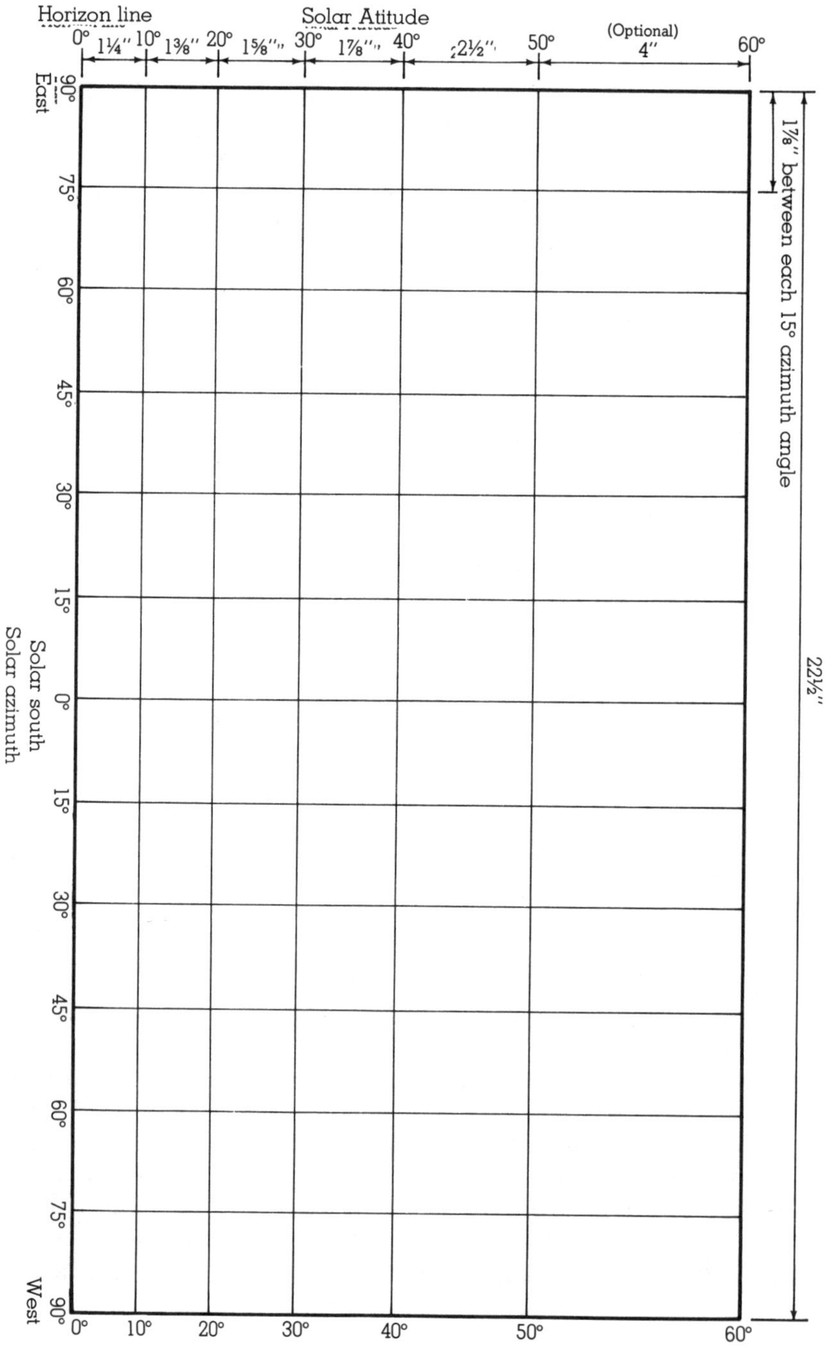

Figure 4–7.
Solar Altitude/Solar Azimuth Graph

line where the curve of the sun paths cross that azimuth. Then, simply connect the dots to form the correct curves. Next, add the time of day lines. Only the portions of the sun's paths between 90° east and 90° west, and up to 60° solar altitude, can be drawn on this shading mask. The sections of the sun's paths not on the mask, mostly those in summer, can easily be imagined to determine if shading is a problem.

5. Find your location on the magnetic Variation Map. Read the number of degrees variation and the direction—east or west. Remember that if you are east of the zero variation line, your magnetic south line is *left* of the solar south line; if you are west of the zero variation line, your magnetic south line is *right* of the solar south line. Draw a dotted vertical line crossing the horizon line representing magnetic south for your location.
6. Trace the sun's paths, solar and magnetic south lines, time of day lines, and horizon line on to the acetate from the prepared graph paper. Leave 2 1/4″ between the bottom of the acetate sheet and the horizon line. Label each of the sun's paths by month, solar and magnetic south, time of day line, horizon line, and east and west.

Figures 4–8 (a-g). Sun paths for northern latitudes

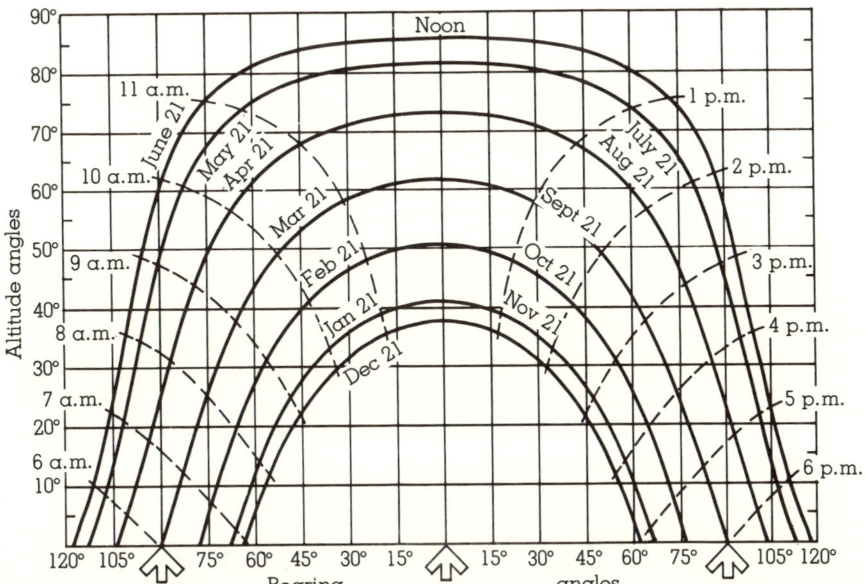

(a) 28° north latitude

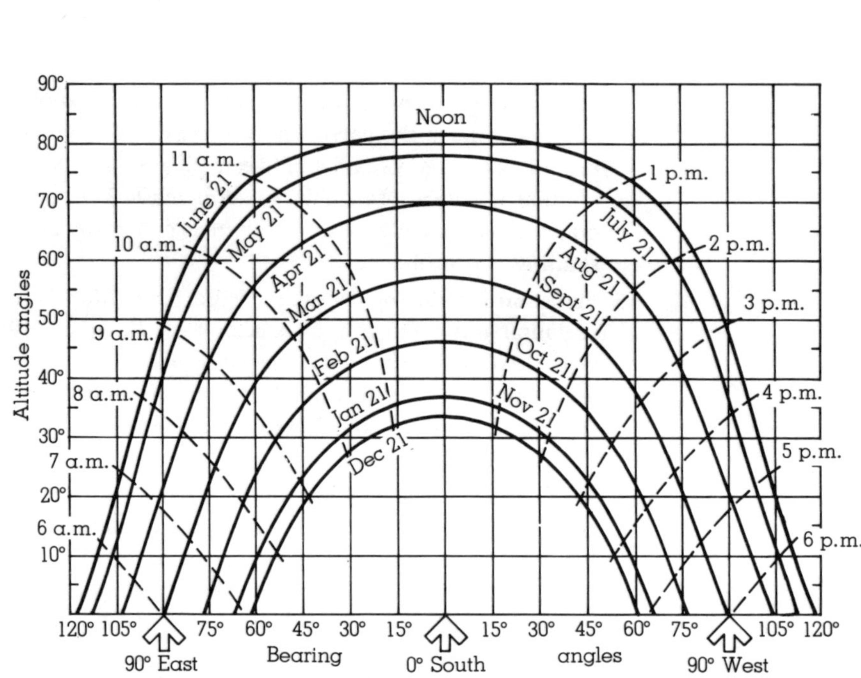

(b) 32° north latitude

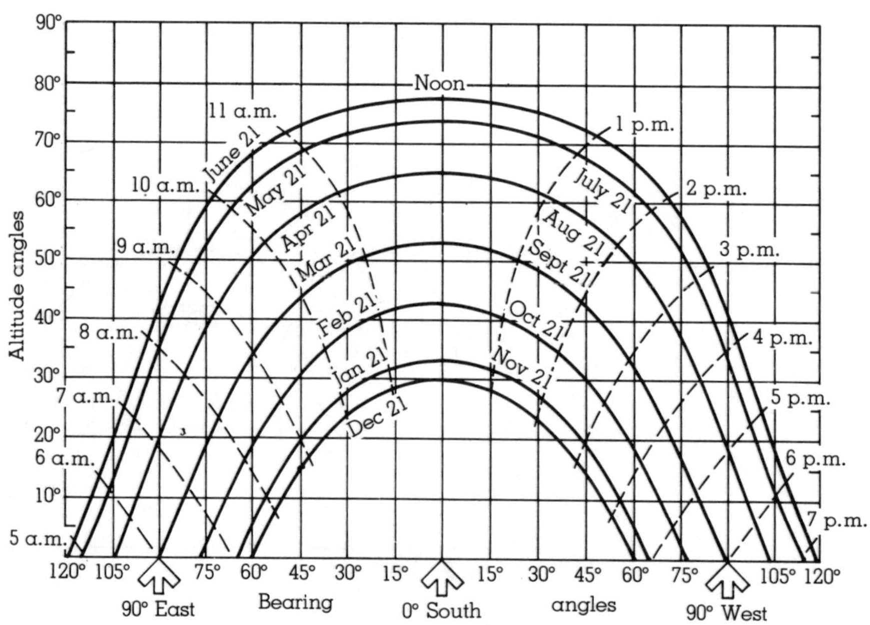

(c) 36° north latitude

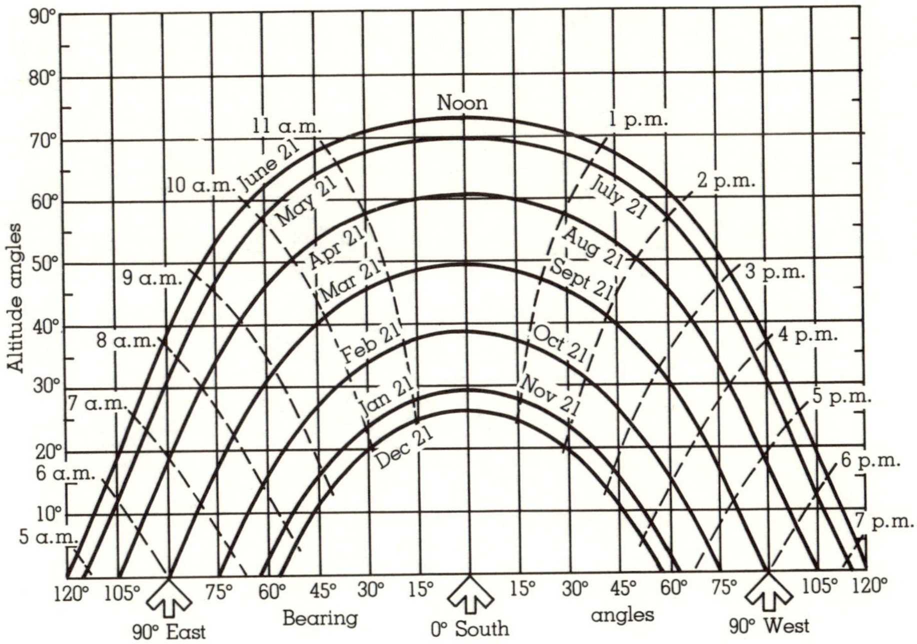

(d) 40° north latitude

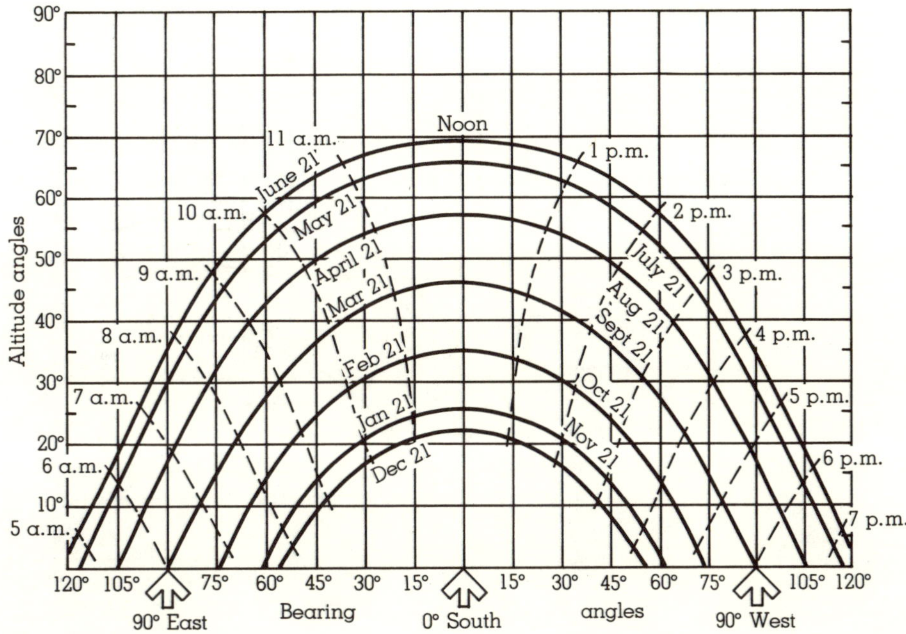

(e) 44° north latitude

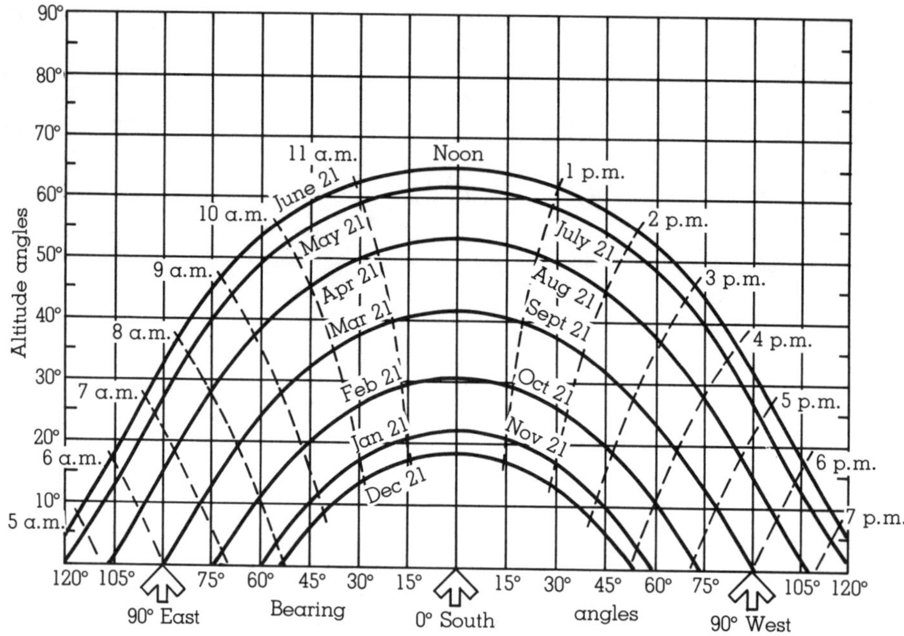

(f) 48° north latitude

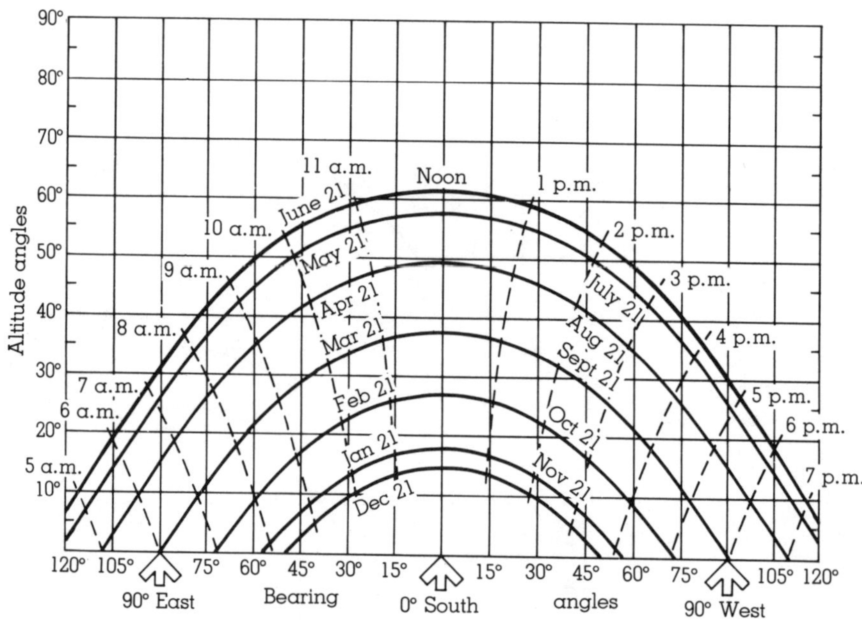

(g) 52° north latitude

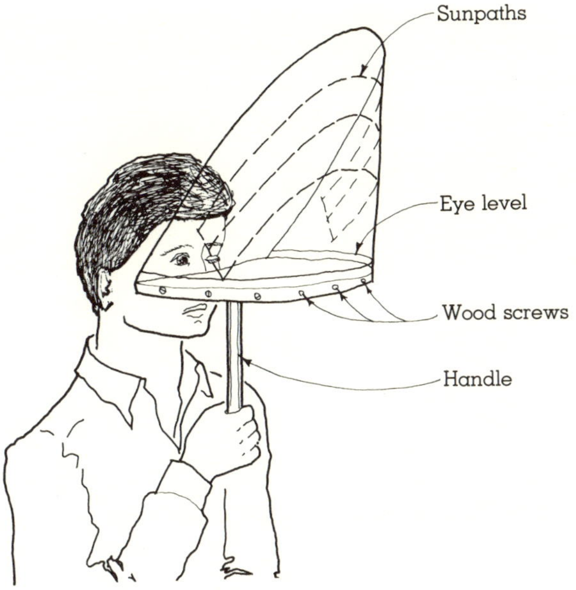

Figure 4–9.
Looking through a solar siting mask.

7. Use the screws, washers, and cardboard to attach the finished acetate sheet to the base. Drill 1/4″ holes through the acetate or Plexiglas sheet, so that the screws do not crack it. Place the cardboard under the washers to add rigidity to the acetate. The horizon line should be 1 1/2″ above the top of the base.
8. Cut away excess acetate about 1/2″ above the highest sun path to add stiffness.

Using the Solar Siting Mask

To determine the amount of shading, stand with your back in line with the potential collector installation. Use a compass to locate magnetic south. Hold the siting mask so that your eye is at the center of the semicircle and level with the horizon line on the mask. Close the other eye. Now aim the magnetic south line on the mask toward magnetic south. Make certain you are holding the base of the mask level. Remember that even small amounts of metal near a compass can disturb its reading, so be sure to check the direction of magnetic south from several positions.

View the sun's path for each month and see when the sun will be shaded. Move to several potential collector sites and compare the amount of shading. Remember that shading of up to 20% of the sun between 9:00 A.M. and 3:00 P.M. is common and acceptable. For further study, sketch the view through the mask on your sun path chart and think about it later.

5 WALL-MOUNTED BATCH HEATER

Figure 5–1.
Before and after installation of wall-mounted batch. This New England house is solar-heated with a four-panel horizontal air-flow active collector, three thermosiphoning air panels, and large south-facing windows. (Designed by Dan Reif; built by Anastas Pollock, Jack Wiren, and Dan Reif. Plans for the three solar house heating systems from *Solar Retrofit: Adding Solar to Your Home,* D. Reif, Brick House Publishing Co., 1981.)

The wall-mounted collector is designed to be attached to the south wall of an existing or new house or apartment. Its long, low shape allows it to fit below windows, on short walls (such as those partially below ground level), or between floors of the building, giving it the flexibility to be suitable for many single and multifamily house designs. This design can also be constructed on flat or low-pitched roofs. The clean lines of the collector, with its steeply pitched top, make it an especially attractive addition. Using the house siding on the collector's sidewalls, the collector becomes an integral part of the house.

The total cost of materials (in 1982) for a quality, high-performance collector with a single solar water tank, 46″ x 76″ triple glazing, and a selective surface is approximately $750. This price includes an average plumbing installation. Tax credits are often available to considerably reduce this cost. In four or five weekends, you should be able to complete this batch design.

DESIGN

The wall-mounted collector design has several important features: 45° sloped glazing, an internal reflector, and a horizontally mounted solar water tank. It also has several design options you will have to select according to the suitability of the south wall, your hot water use, and your budget.

The 45° glazing angle was selected because it provides excellent year-round solar performance for the entire central and northern U.S. and southern and central Canada. For the southern U.S. (latitudes less than 30°), conventional guidelines suggest that collector glazing be sloped more horizontally to achieve better summer performance. But in these warm climates, summer performance is probably more than adequate even with the 45° sloped glazing. In more northern latitudes (latitudes greater than 60°), collector glazing can be sloped more vertically for better winter performance—but only if the collector is going to be used through the coldest part of the year. Otherwise, the 45° angle is fine.

The 45° angle also makes efficient use of materials and simplifies construction.

The collector's interior surfaces are lined with foil-covered insulation to reflect additional sunlight onto the solar water tank. Although these reflective surfaces are flat and therefore do not focus the light onto the tank the way curved reflectors do, the tank is so large in comparison to the reflectors that almost all of the reflected light strikes the tank. Since these reflectors are inside the glazing, they do not require maintenance.

The horizontal placement of the solar water tank was selected to give the collector its compact, low profile suitable to the available space and esthetics of most existing and new homes. The drawback of the horizontal placement is that better temperature stratification occurs when the tank is oriented more vertically. This disadvantage has only minor impact on performance. To increase temperature stratification in horizontal systems, separate the inlet and outlet water pipes as much as possible, and reduce the mixing of the cold inlet water with the tank's warm water.

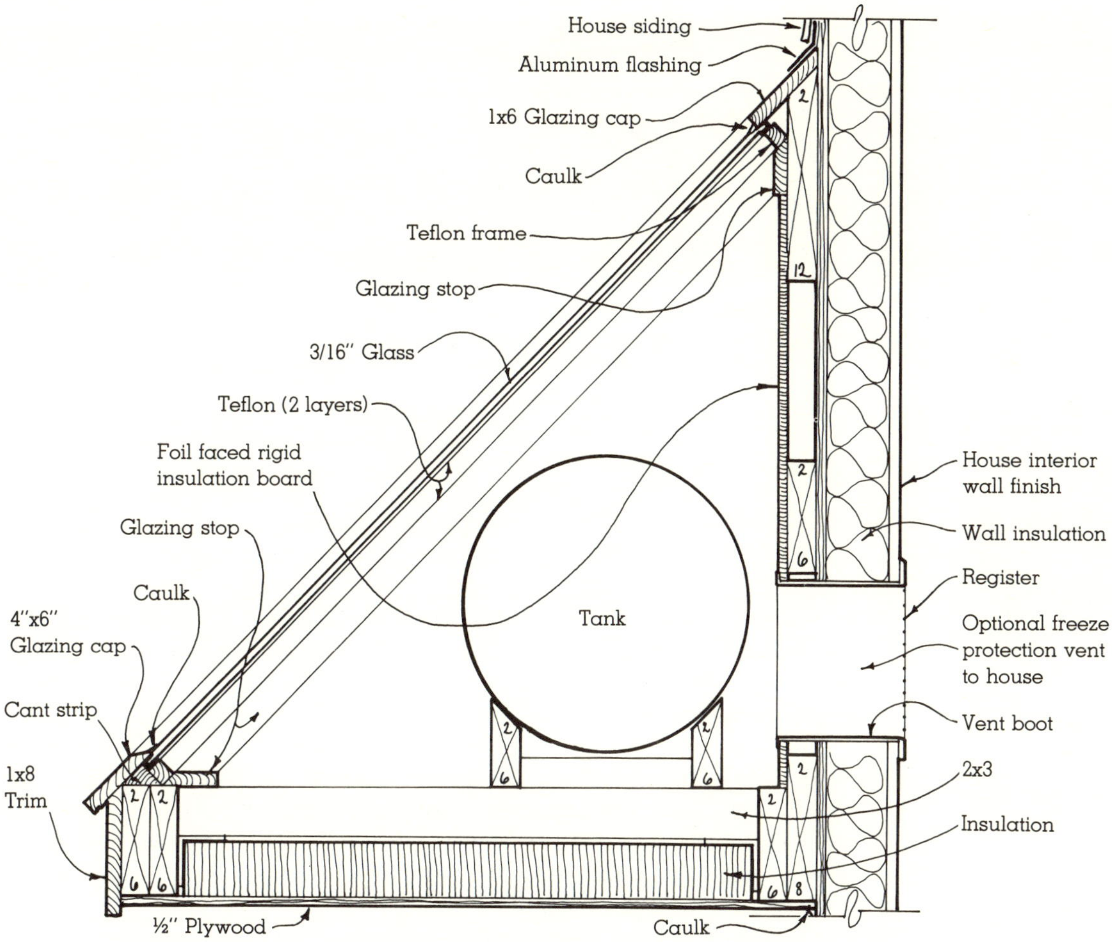

Figure 5–2. Side view of wall-mounted batch water heater. *Note*: Locate the tank approximately 1 1/2″ from the inner glazing.

Design options for the wall-mounted collector include: the number and size of the solar water tank(s), the number of layers and type of glazing material(s), and the selection of material to cover the solar water tank(s).

This collector can be constructed with one or more solar water tanks. With one 46″ x 76″ glazing unit providing a net glazing area of 24 sq. ft., a water tank of between 36 and 54 gal. is recommended. Two glazing units can be built side by side providing space for a second tank. Each tank should be between 12″ and 16″ in diameter. With the 76″-long glazing, use a tank approximately 5′ long (5′10″ is the maximum length). When using several tanks, plumb them in series to increase the temperature stratification of the solar-heated water.

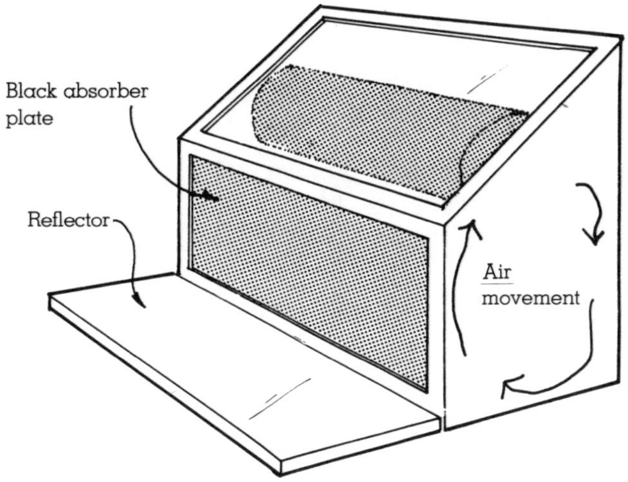

Figure 5–3.
Proposed thermosiphoning air panel (TAP) booster for batch.

Glazing for the wall-mounted collector should consist of two or three layers. My preference is one layer of glass as the outer glazing and two layers of Teflon film for the inner glazing. Use the recommended glazings on pages 50 through 53 to assist you in your selection.

The solar water tank(s) can be painted flat black or covered with a selective surface. The selective surface adds approximately $60 per 40- to 60-gal. tank to the cost of the collector, but helps significantly in retaining warm water throughout cool nights. The selective surface combined with double glazing has similar performance to flat black paint and triple glazing. When my budget can afford it, I use the best—triple glazing and a selective surface.

Another option for cool and cold climates is the use of an operable register or fan-powered vents between the collector and house. In cool weather, the register allows warm house air into the collector as a means of freeze protection. During long periods of below-freezing temperatures and cloudy weather, the solar tank is drained and the fan blows hot collector air into the house. *This option allows the wall-mounted collector to provide efficient year-round service even in very cold climates.*

LOCATION

Determining the exact location of your batch on the south wall is a major decision affecting solar performance, ease of construction, and esthetics.

First, find a very sunny location for the collector. Shading is a major cause of poor performance. If there is any question regarding the amount of shading or solar orientation, do a thorough site evaluation.

Figure 5–4.
To locate the wall studs, remove the house siding, leaving the sheathing in place. The sheathing will be nailed to the studs.

The construction sequence in this chapter is for a collector with one 46″ x 76″ (standard glass size) glazing unit. The collector needs a flat, uninterrupted wall space at least 44 1/2″ high by 7′, 7 1/4″ long. A two-unit collector can also be built from these plans and needs the same height, but a space 14′, 5 3/4″ long. Other size glazing units and the use of other glazing materials can easily be adapted from these plans.

This wall-mounted collector is designed to be attached to standard 2″ x 4″ or 2″ x 6″ wood-frame construction. With the use of masonry bolts (Hilti fasteners are recommended), the collector can be mounted on brick or concrete walls. A modified wall-mounting system or several foundation posts can be used to attach the collector to any wall surface.

Ideally, the batch should be located so that the collector's sidewalls are positioned directly over studs in the house wall. This can be done in new construction. It provides an easy way to secure the collector sidewalls to the house. If the collector sidewalls are not located over wall studs, which often occurs in retrofits, then the collector's three horizontal mounting supports provide the secure base on which to mount the sidewalls. Wall studs can sometimes be located by finding the vertical rows of nails in the house siding. Otherwise, you will have to locate the studs after some of the exterior siding has been removed, or locate them from the interior of the house. If you remove some of the siding in the center of the collector mounting area, you can locate the exact position of the studs by noting the nails in the house sheathing. Then you can determine the exact position for the collector's sidewalls.

Plumbing from the auxiliary water heater to the wall-mounted collector can often be installed in a simple, straightforward manner with minimum risk of

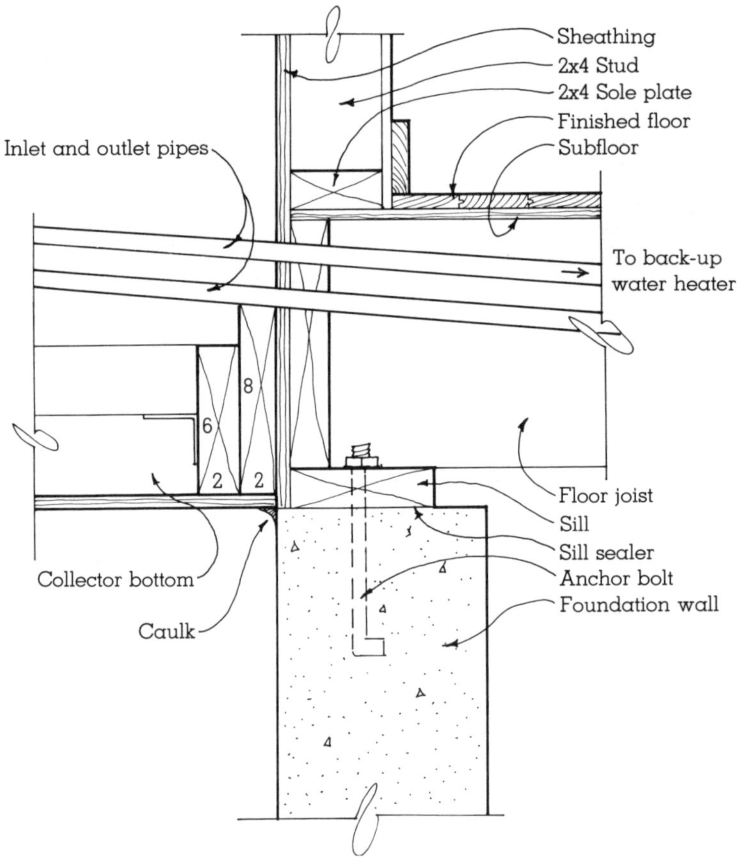

Figure 5–5.
Placing the bottom of the collector along the bottom of the house sill can simplify the plumbing connection to the house.

winter freezing. This can usually be accomplished by placing the bottom of the collector level with the bottom of the house sill plate, and installing the inlet and outlet water pipes between the house floor joists on the warm side of the insulation. The pipes should pass just above the top of the collector base and at a slight downward slope into the house. This approach eliminates the need to install plumbing in the house interior walls, a difficult task in existing houses.

If you decide to use either an operable register for freeze protection or fan-powered solar space heating, you will have to determine the location of the inlet and outlet vent registers. The register must go directly through the adjoining collector/house wall. Natural air movement is too weak to travel horizontally through long air ducts. For solar space heating, a fan is needed to efficiently move the air from the collector into the house. Fortunately, the fan can deliver warm collector air through ducts to anywhere in the house.

Figure 5–6.
Wall-mounted batch based on plans presented in this chapter, combined with fan-assisted thermosiphoning air panels (the dark panels on the wall) for solar house heating.

PERFORMANCE

Several wall-mounted collectors have been constructed and monitored in New England and Pennsylvania.

Anastas Pollack, Jack Wiren, and myself constructed a batch heater in Connecticut using the plans in this chapter (figure 5–6). The system consisted of a single 40-gal. tank covered with a selective surface foil and 24 sq. ft. of triple glazing—low-iron outer glazing and two layers of Teflon inner glazing. The six-person household uses 60 to 100 gal. of hot water each day. Because the daily hot water demand far exceeds the solar water tank supply, the system was plumbed to serve as a preheater to a 40-gal. gas-fired water heater. A thermosiphoning vent and Frost-Tex heat tape were used to provide freeze protection in the late fall. During the winter the solar tank is drained and a 65 CFM thermostatically controlled fan is used to move solar-heated air into the house.

Year-round performance has been excellent. In the late spring and early fall, with outdoor temperatures in the 40s and 50s at night and 60s and 70s during the day, the batch provides approximately 50% solar water heating. After 3 or more hours of bright sun, the solar water temperature often exceeds 100°F—even though in this active household hot water is used throughout the day. On sunny winter days (with the solar tank drained) the collector supplies the majority of heat needed during the day by a 12′ x 20′ room.

Nature's Way Energy Systems in Keene, New Hampshire, has constructed several similar collectors using triple glazing, selective surface, and vents to allow house-heated air into the collector to provide freeze protection (see page 20). Nature's Way reported that neither the solar tank nor pipes froze despite −27°F temperatures! (That's right, −27°F.) The pipes in these collectors are kept short by exiting the collector enclosure as soon as possible. R-8 pipe insulation is used, but *no* electric heat tape is provided. One advantage that these batches had was that they were not in use and therefore did not have cold water put into them at night.

Conservation Concepts in Hummelstown, Pennsylvania, has been constructing wall-mounted collectors for several years (see page 35). Their collector uses two 40-gal. tanks plumbed in series, 40 sq. ft. of double fiberglass-reinforced plastic glazing, and flat black paint.

In February and early March, with daytime outdoor temperatures between 20°F and 40°F, the tanks' average daytime temperature was between 90°F and 110°F. During sunny periods the temperature peaked at almost 140°F. Morning temperatures were usually near 60°F, and rarely dropped below the normal well water temperature of 50°F.

By late May, 100% solar water heating was attainable with some adjustment of hot water use. By mid-June the solar water heater supplied all of the normal hot water needs. Average summer water temperatures ranged between 125°F and 150°F during the day, with peak temperatures approaching 170°F.*

CONSTRUCTION

Safety

Building a quality batch water heater requires good plumbing and carpentry skills. You should know the proper procedures for performing the work described and be familiar with the potential hazards. Safety is based on knowledge, skill, and an attitude of care and concern.

- Keeping your construction site neat and orderly makes the construction process both easier and safer.
- Use the correct tools and keep them in good working order. Poorly functioning or improper tools subject you to unnecessary dangers.
- Wear safety glasses when using power tools or whenever the work involves hazard to your eyes. Also, wear shoes with thick soles to protect your feet from protruding nails.
- Do not place tools overhead on ladders or anywhere else where they may slip off and cause injury.
- Lift heavy objects with the strength of your legs—not with your arms and back. Always stand close to the load and keep your back as vertical as possible. When the load is extra heavy or bulky, get help from friends.

*"Passive Solar Workshops in the Shadow of Three Mile Island," Proceedings of the Fifth Passive Solar Conference, 1980, by Mark Whitmoyer and Paul Swartz.

- Keep a bucket of water and a fire extinguisher handy when soldering. For soldering, the water pipes are drained and it will be difficult to get water quickly in case of fire.
- Be especially careful when working on a roof. Roof work requires special skills and sometimes special tools—scaffolding, chicken ladder, extension ladder. Do not work on a roof unless you are completely prepared for it. Do not work if the weather is windy or wet. Wear rubber-soled shoes. Watch out for electrical wires. And, make sure you have plenty of assistance when lifting the solar water tank into position.

Materials List for a Single-Tank Unit with 46″ x 76″ Glazing

Part	*Quantity*	*Description*
horizontal mounting supports	1	8′ — 2″ x 6″
	1	8′ — 2″ x 8″
	1	8′ — 2″ x 12″
		3/8 diameter lag bolts and washers (see text for sizes and amounts)
sidewalls	1	1/2″ CDX plywood
	4	8′ — 2″ x 6″
	14	4″ or 6″ 3/8″-diameter lag bolts and washers (see text for size)
	4	8″ 3/8″-diameter lag bolts and washers
	10 sq. ft.	6″ fiberglass insulation
	10 sq. ft.	6 mil polyethylene (vapor barrier)
	1	1/2″ foil-faced rigid insulation board
bottom	1	1/2″ CDX plywood
	1	8′ — 2″ x 6″
	2	10′ — 2″ x 6″
	1	12′ — 2″ x 3″
	8	3″ metal ell brackets with 1 1/2″ wood screws
		foil-faced rigid insulation board (R-20 minimum)
tank support frame	1	14′ — 2″ x 6″
selective surface foil		see text for size (optional)
Frost-Tex heat tape		optional
pipe insulation		
aluminum flashing	5′	10″ wide (for vent boots)
glazing stops	2	8′ — 1″ x 2″

Part	Quantity	Description
Teflon frame	1	8′ — 1″ x 6″ clear (no knots)
Teflon	1	50″ x 14′
glass	1	46″ x 76″ textured low-iron glass
	2	2″ long 1/2″ x 1/4″ neoprene setting blocks
	8	2″ long 1/2″ x 1/8″ neoprene spacers
glazing caps	2	8′ — 1″ x 6″
	1	8′ — 1″ x 10″
trim	2	8′ — 1″ x 4″
	1	8′ — 1″ x 8″
aluminum flashing	8′	6″ wide
nails	5 lbs.	16d galvanized (hot-dipped galvanized)
	2 lbs.	6d galvanized (hot-dipped galvanized)
wood screws	60	1 1/2″ — #6
aluminum foil duct tape	1 roll	
flat black paint	12 oz.	spray can
exterior glue	1 qt.	
wood preservative	1 qt.	
white paint	1 qt.	
felt weatherstripping	25′	1/4″ x 1/2″
clear silicone or urethane caulk	2	10-oz. tubes

1. Preparing the House Wall

Using a level and tape measure, draw the outline of the collector on the south wall:

7′, 7 1/4″ long x 3′, 6 1/2″ high for a one-tank, 46″ x 76″ glass unit with 2″ x 6″ sidewalls, or

14′, 5 3/4″ long x 3′, 6 1/2″ high for a two-tank, two 46″ x 76″ glass units with 2″ x 6″ sidewalls.

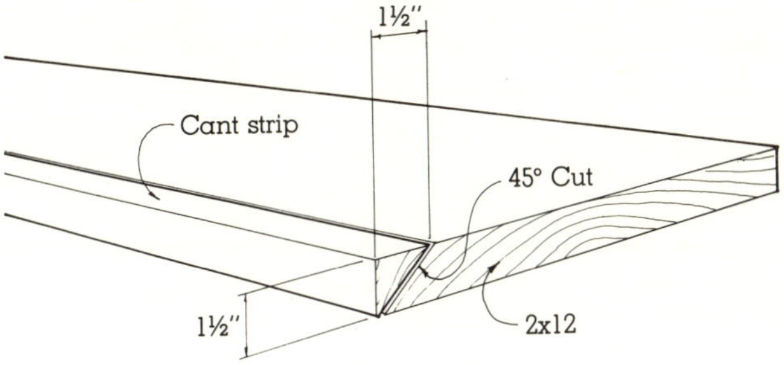

Figure 5–7.
2″ x 12″ horizontal mounting support and cant strip.

Check the squareness of the outline by making sure that the diagonal measurements of the outline are equal. These dimensions allow 1/4″ clearance on the top-to-bottom and side-to-side measurements.

Remove the house siding from the collector mounting area. Leave the sheathing but remove all building and tar paper covering the sheathing. Work carefully so that you can save enough of the house siding to use on the collector sidewalls, thus making a perfect match between the collector and the house. A good way to cut clapboard or shingled siding is to temporarily nail a flat, straight piece of wood over the siding and use this wood strip as a smooth surface on which you can guide a circular saw. Set the saw blade depth so that it cuts only the siding and not the sheathing.

Caution: Wear goggles since there might be hidden nails.

When removing the siding along the wall where the top of the collector will be, visualize how the flashing will be placed so that it can fit under the house siding and then over the top edge of the collector. Pry this house siding slightly upward to accommodate the flashing.

2. Securing the Horizontal Mounting Supports

Now you are ready to start constructing the collector frame. First cut 8′ long pieces of 2″ x 6″, 2″ x 8″, and 2″ x 12″ to 7′, 4 1/2″ long. These are the *horizontal mounting supports.* Use a circular saw to cut a 45° angle along the top edge of the 2″ x 12″. The cut-off triangular piece measuring 1 1/2″ x 1 1/2″ x 2 1/8″ (7′, 4 1/2″ long) is the *cant strip* you will need later.

The horizontal mounting supports secure the collector to the house wall. These supports must be carefully mounted to ensure that all connections are very solid! A single-tank, 40-gal. system weighs approximately 600 lb.

The horizontal mounting supports are located 1 3/8″ from the vertical edges of the collector mounting area. Before securing the mounting supports, check their vertical alignment with a level. The 1 3/8″ spaces are for the 1/2″ plywood on the collector sidewall, 3/4″ exterior trim, and caulk.

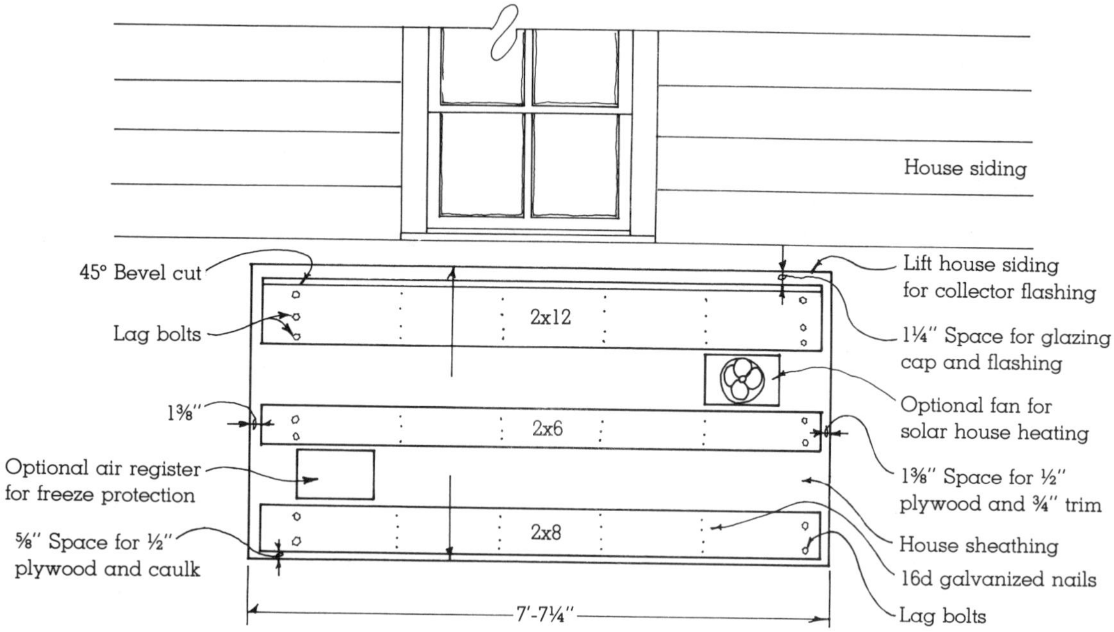

Figure 5–8. Installation of horizontal mounting supports.

The 2'' x 12'' mounting support is placed with its beveled edge on top and 1 1/4'' below the top edge of the collector mounting area. The 1 1/4'' space is for the glazing cap and flashing. The bottom 2'' x 8'' mounting support is placed 5/8'' above the bottom of the mounting area leaving space for 1/2'' plywood on the collector bottom and for caulk.

Each of the three horizontal mounting supports is secured to the house wall studs and sheathing with 16d galvanized nails and 3/8'' diameter lag bolts. The lag bolts are only used at the ends of the mounting supports where the support is directly over the last wall stud at each end. Nails are used to fasten the mounting supports to the other studs. Drive the nails in at opposing angles to provide additional holding power. The length of the bolts depends on whether or not the end 5'' of the mounting support are located directly over a house wall stud.

If the end 5'' are located over a stud, then 6''-long lag bolts are used later in step 3 to screw through the collector sidewall and horizontal mounting support and into the house sheathing and stud. If the end 5'' of the mounting support are not over a stud, 5''-long lag bolts are used to screw through the mounting support, house sheathing and stud; later 4''-long lag bolts are screwed through the collector sidewall and into the mounting supports. Use two lag bolts for each connection

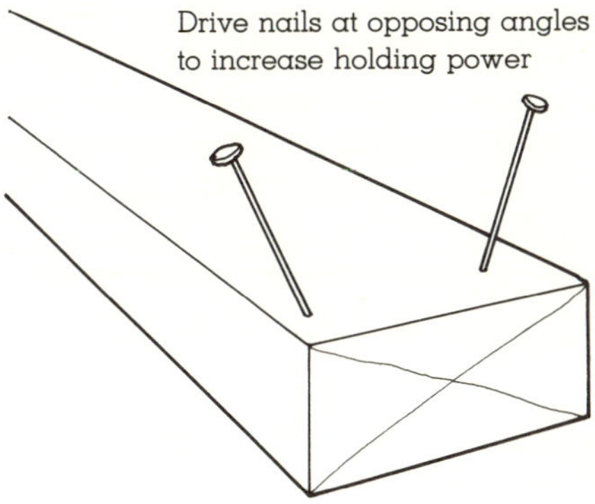

Figure 5–9.
Drive nails at opposing angles to increase the holding power of the horizontal supports.

Figure 5–10.
Horizontal mounting supports.

through the 2″ x 6″ and 2″ x 8″ mounting supports, and three lag bolts at each 2″ x 12″ connection.

Before installing any lag bolts you must find the *exact* location of the wall studs you are going to bolt into. If you miss the center of the stud, you can split the stud, resulting in a very weak connection. Also, you *must* check that there are no wires or pipes passing through the stud where you are piercing it with the lag bolts. To locate the exact position of the stud, drive several nails through the sheathing until you strike the edge of the stud. Now drill a 1″ diameter "peep hole" through the sheathing adjacent to the stud. Use a screwdriver to probe into the exposed wall cavity and check for wires and pipes. Check each stud before drilling the bolt holes.

Now, drill holes large enough to countersink the washer, then a 1/4″ diameter hole for the 3/8″ diameter lag bolt. Screw the bolt in until its head is flush with the wood.

3. Cutting Openings in the House Wall for Plumbing

Cut a hole approximately 3'' high and 6'' wide through the house wall for the inlet and outlet water pipes and the insulation around them. The bottom of the hole must be within the upper 2'' of the lower collector mounting support so that the collector bottom can be mounted without interference. Remember that both the inlet and outlet pipes must be sloped slightly downward toward the house so that the pipes within the collector can be completely drained when necessary. To minimize the length of pipes inside the collector, place the hole near the end of the collector where the pipes will enter the solar water tank.

4. Freeze Protection

In cold climates Frost-Tex electric heat tape (made by RayChem, Redwood City, California) is recommended to protect the pipes from freezing. This type of heat tape gets hotter as the air temperature gets colder, and it does not use a thermostatic switch, which could fail.

Another option for freeze protection is to allow warm house air to circulate through the collector during cold weather. Operable air registers can be installed through the house wall into the bottom of the collector. Since the pipes are more prone to freezing than the tank, locate the register near the solar tank's plumbing. Two registers can be used if there are pipes on both sides of the collector. The electric heat tape can also be used as insurance.

When installing an air register a vent boot should be used to enclose the passage between the house and collector preventing warm air from infiltrating into the house wall. To install an 8'' x 14'' operable register and vent boot, first cut a 14 1/4'' wide by 8 1/4'' high opening in the house sheathing. Next, nail wood blocks between the house sheathing and interior wall surface to provide a surface onto which you can later mount the vent boot and register. With a keyhole saw or utility knife you can now remove the interior wall surface within the vent opening.

The vent boot is made from 10''-wide aluminum flashing. Bend and cut the flashing so that it fits snugly within the 8 1/4'' x 14 1/4'' opening. The vent boot should project at least 2 1/4'' into the collector to provide for the thickness of the collector's rear wall, and should have an approximately 1/4'' flange, which is bent flush with the interior of the house wall. Seams in the vent boot should be sealed with caulk. Secure the boot by nailing it to the wood blocking.

Note: Allowing warm, moist house air into the collector can cause condensation on the cool tank and glazing. To protect the collector from this moisture, all wood within the collector must be treated with several coats of wood preservative.

5. Optional Freeze Protection and Solar House Heating

A fail-safe option for freeze protection during the coldest months of the year is to drain the solar water heater tank and all exterior plumbing. This is recommended for cloudy months of more than 1,000 degree days. Degree days per month for cities throughout the United States are listed in the appendix. *Fortunately, this solar water heater design can easily be adapted as a very effective daytime solar space*

Figure 5–11.
Fan-powered vent for optional solar house heating.

heater. This provides year-round use of your collector, giving you the most solar energy in return for the time and money invested.

Note: Cover the glazing to prevent overheating the collector when the solar water tank is drained and solar space heating is not in use.

To use the collector as a solar space heater, install a small fan at the top of the wall between the collector and house. Place a cool-air return vent along the bottom of the collector/house wall. A thermostat should also be installed to control the operation of the fan. Len Meserve, Jr. of Nature's Way Energy Systems first showed me this important design innovation.

To move the air efficiently from the collector to the house, the fan size should be rated to move between 2 and 3 cubic feet per minute (CFM) for each square foot of collector glazing. For example, a 46″ x 76″ glazing unit has approximately 24 sq. ft., and therefore should have a fan rated for 48 to 72 CFM.

The cool-air return vent should be located diagonally opposite the collector fan. Use a register measuring approximately 8″ x 14″.

The thermostat should turn the fan on whenever the collector temperature is several degrees warmer than the house, and off when the collector is near or below the house temperature. Snap-disc and adjustable bimetal thermostats are inexpensive and available from heating, venting, and air conditioning suppliers. Note that you need a *cooling* type thermostat that turns the fan *on* when the temperature *rises*. Also, the thermostat must be able to withstand temperatures of approximately 200°F. (Thermostats and fans are available through mail order from Solar Components Corporation, P.O. Box 237, Manchester, NH 03105 and wholesale to the trade from Grainger's, 5959 West Howard Street, Chicago, IL 60648.)

With the addition of a second thermostat control, this fan system can also be used for freeze protection. Install a *heating* type thermostat that will turn the fan on

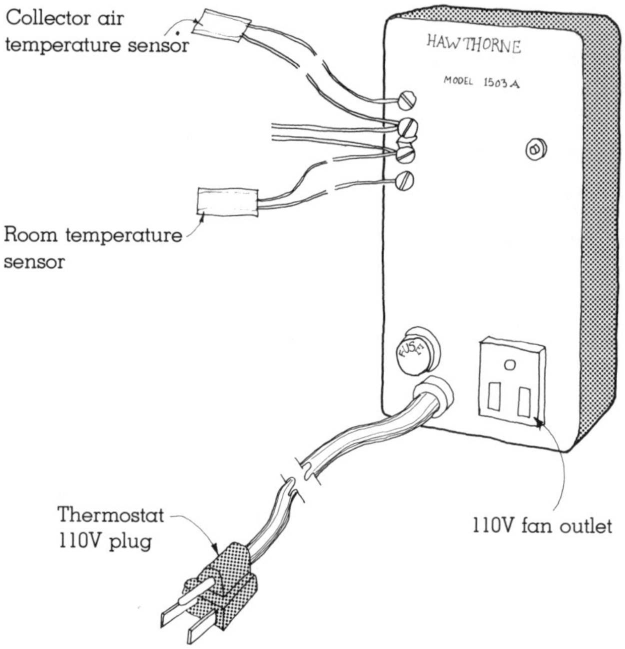

Figure 5–12.
Hi-Square 1503-A differential thermostat.

whenever the collector air temperature is below 40°F. As with any electrically operated freeze protection, winter power outages can be disastrous.

Note: Some thermostats include both cooling and heating functions.

If you are not experienced with electrical work and do not want to hire an electrician, you can use a thermostat such as the Hi-Square 1503-A. This thermostat plugs into an existing wall outlet, then you simply plug the fan into the outlet mounted on the thermostat. This differential thermostat turns the fan on whenever the collector is 16°F warmer than the house. It turns the fan off when the collector temperature is less than 3°F warmer than the house.

Another nice feature of this thermostat is that an additional temperature sensor can be added to turn the fan on when the collector temperature is below 38°F. This can be used as an effective method of freeze protection.

One disadvantage of the Hi-Square thermostat ($70) is that it costs much more than the snap-disc thermostat ($8).

Note: When the weather warms and you reactivate your solar water heater, remember to disconnect the fan and completely stuff all the vent openings with insulation.

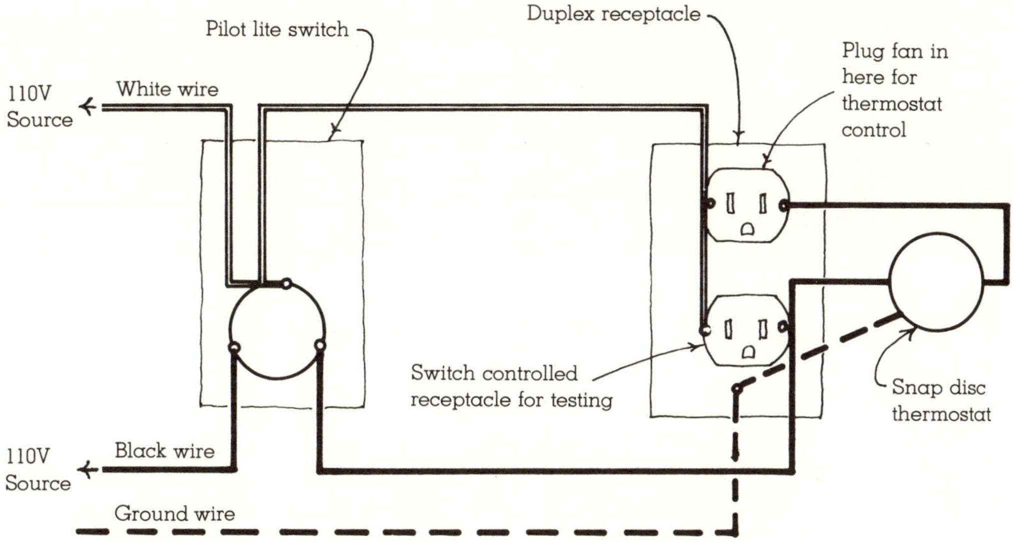

Figure 5–13. Electrical diagram for solar house heating fan with snap-disc thermostat.

To install a solar space heating fan and snap-disc thermostat, study figure 5–13 and follow these steps.

1. The duplex receptacle shown above is wired so that each outlet functions separately. To separate the outlets, break the metal tab between the screws on the black wire side only. The top outlet is controlled only by the pilot switch. This outlet is used for testing the fan system. The bottom outlet is controlled by both the thermostat and switch.
2. Be sure that the electrical source is not controlled by a switch that might occasionally be turned off.
3. Connect the white wire to the switch only if a pilot light is desired. Otherwise, only break the black wire at the switch.
4. Ground the switch, receptacle, thermostat, and fan.
5. Use high-temperature wire between the thermostat and receptacle.
6. For thermostatically controlled fan use, simply plug the fan into the lower outlet.

Caution: Electricity can be dangerous! Do not install your own wiring unless you know what you are doing. Always follow local building codes.

For more information, see *Wiring Simplified* by H.P. Richter (Minneapolis, Minn.: Park Publishing, 1978).

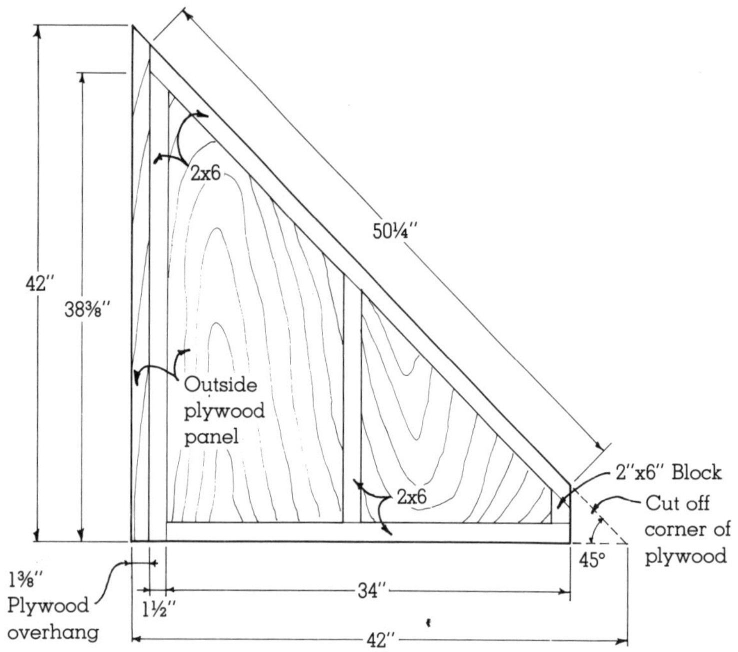

Figure 5–14.
Sidewall assembly.

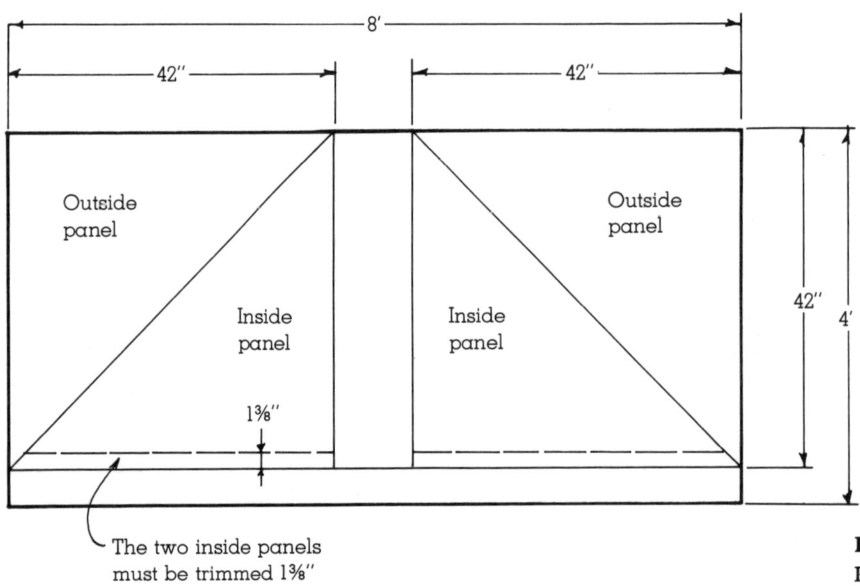

Figure 5–15.
Plywood cutting pattern for collector sidewalls.

Figure 5–16.
Collector sidewalls.

6. Constructing and Mounting the Collector Sidewalls

The next step is to prefabricate the collector sidewalls. Assemble these components on flat ground.

Each of the two sidewalls uses five pieces of 2″ x 6″ and two triangular pieces of 1/2″ exterior grade plywood. The two sidewalls are identical except that the outer piece of plywood is nailed on the opposite side of each section. The four triangular plywood pieces for both sidewalls can be cut from one 4′ x 8′ sheet of 1/2″ CDX plywood. The size of the plywood and 2″ x 6″ pieces are shown in figure 5–15. Label each piece as you cut it to avoid confusion when assembling. Also, label the right and left sidewall.

Use 6d galvanized nails and exterior-grade glue to fasten the 2 x 6s to the plywood. The bottom and sloped 2 x 6s should be flush with the edges of the plywood, and the back 2 x 6 should be 1 3/8″ from the back of the plywood. Use 16d galvanized nails to secure the 2 x 6s together.

Now mount the sidewalls to the ends of the horizontal mounting supports. Use 4″ or 6″ x 3/8″ diameter lag bolts as described in step 2. Two additional 8″ x 3/8″ diameter lag bolts are then screwed through each sloped 2″ x 6″ into the house wall at a 45° angle. The 8″ bolts must be countersunk so that the heads of the bolts are below the surface of the 2″ x 6″.

Completely fill the inside of the sidewalls with insulation, then cover the insulation with a vapor barrier such as 6 mil polyethylene. Use 6d galvanized nails and glue to secure the inside plywood to the sidewalls.

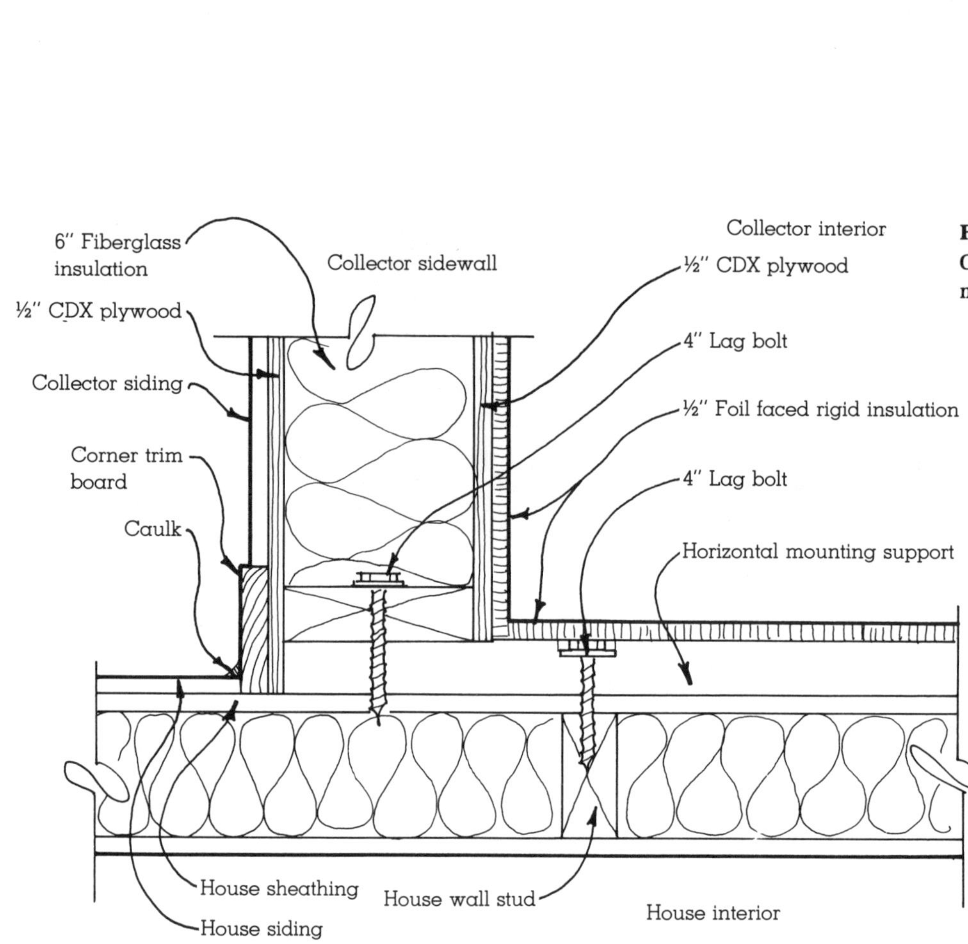

Figure 5–17. Collector sidewall mounting detail.

7. Constructing and Mounting the Collector Bottom

The collector bottom is made of a 2″ x 6″ frame, 2″ x 3″ joists, a rectangular sheet of 1/2″ exterior grade plywood bottom, and aluminum foil-faced rigid foam insulation board. The bottom is assembled so that the plywood, insulation, and joists can be removed after the collector has been completed. This provides easy access to the plumbing, tank, and inner glazing in case servicing is needed after completion of the system.

Use 16d galvanized nails to assemble the 2″ x 6″ bottom frame. Now use 1 1/2″ wood screws to secure metal ell brackets to the 2″ x 6″ and 2″ x 3″ joists. Space the four 2 x 3s so that the solar tank will be supported by all of them.

Nail the cant strip flush with the front edge of the 2″ x 6″ furthest from the house wall. The cant strip is the triangular piece of wood that was cut off from the top horizontal mounting support in step 2. Since the collector bottom is exposed to direct sunlight, it should be painted with several coats of white, exterior paint. Also, avoid using pine since it may outgas at high temperatures.

Now secure the collector bottom to the sidewalls and lower horizontal mounting support with 16d galvanized nails.

Cut the foil-faced rigid insulation board to fit snugly within the collector bottom. Also, cut a piece of 1/2″ foil-faced foam board to mount against the back of the collector over the horizontal mounting supports.

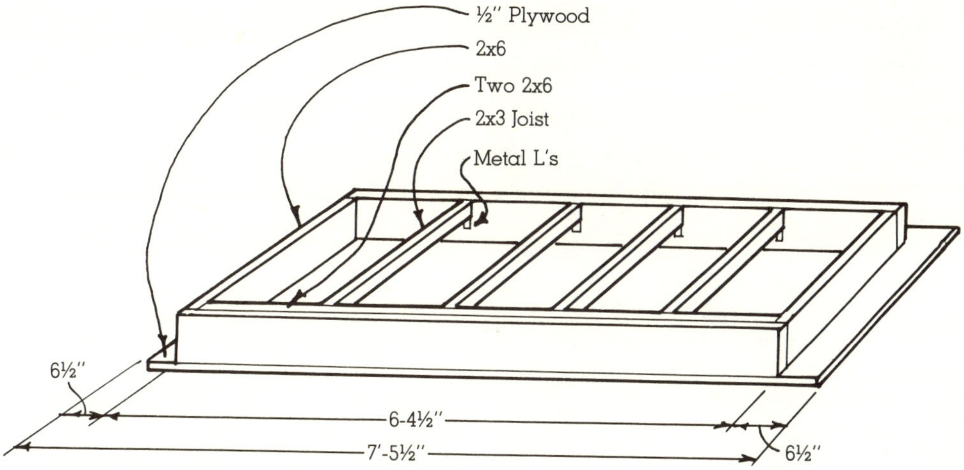

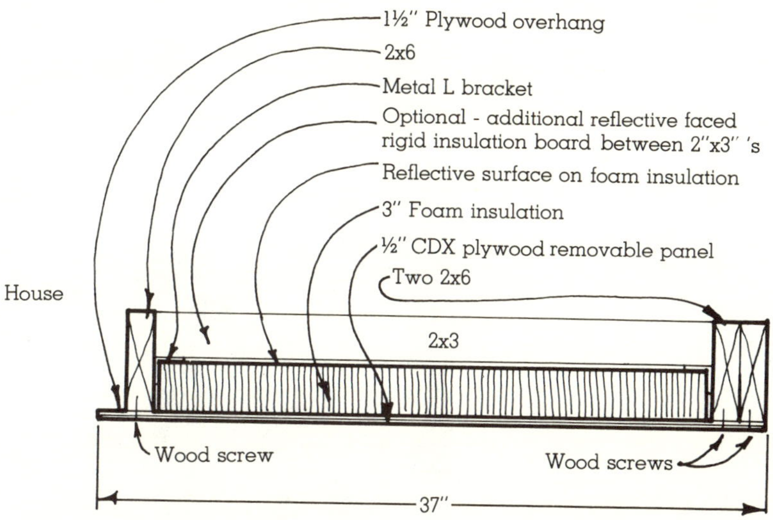

Figure 5–18. Collector bottom with removable servicing panel.

Note: Leave space along the upper horizontal mounting support for the upper glazing stop.)

Since the insulation board is the finished interior surface, make all cuts accurately. For any vent openings, complete the vent boots by bending the aluminum flashing to form a flange flush with the insulation board. Wrap all exposed edges of the insulation board with aluminum foil duct tape. The foam insulation bottom should have a minimum R-value of 20. After the insulation has been completed, screw the plywood bottom into the collector bottom, sidewalls, and horizontal mounting support.

Figure 5–19.
Placing foil-faced insulation in the collector.

Figure 5–20.
Test the strength of the collector before proceeding.

The collector frame should now be strong enough to support 500 pounds. If for any reason the frame is not strong enough, you must add additional supports. You can construct a triangular brace connecting the bottom of the collector and the house, or you can add two foundation supports under the front of the collector. These supports can be made from pressure-treated wood posts or small concrete piers. These supports must be approved by the local building codes and extend below the frost line. Check with your local building inspector for details.

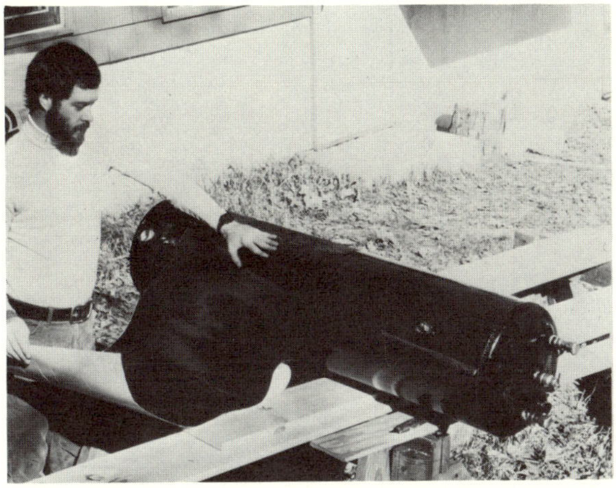

Figure 5–21.
Applying the selective surface foil to the tank.

8. Installing the Tank

Before placing the tank into the collector you have to prepare the tank's surface, construct a tank support frame, and mount the necessary fittings onto the tank.

The tank can be either painted or covered with a selective surface foil. To prepare the tank for painting, simply clean it thoroughly with a strong detergent to remove grease and oil, then rinse it. Applying a selective surface foil takes a bit more work.

First, the tank should be lightly sanded until its surface is smooth. Next, clean the tank with detergent and rinse it. Since the foil is difficult to apply to the rounded ends of the tank and on the tank's fittings, paint the ends and area around the fittings flat black.

The selective surface foil I have used is called Sunsponge and is made by Berry Solar Products (P.O. Box 327, Edison, NJ 08817). The foil comes on a 24″-wide roll and is available with an adhesive backing that makes it easy to apply. Cut a strip of selective surface foil long enough to wrap around the diameter of the tank plus a 1″ overlap. Start peeling the protective sheet off the adhesive backing, then starting at one end of the tank, apply the foil around the tank diameter. Work carefully to keep the foil smooth and wrinkle-free. Keep the clear protective film in place over the selective surface until the tank is ready to be placed in the collector. An assistant will certainly make this task easier. Use a soft rag or wallpaper roller to firmly press the foil onto the tank and remove air bubbles. Prick any remaining bubbles with a pin and press them smooth. Apply one or two additional wraps of foil to cover the curved sides of the tank. Each successive sheet of foil should overlap the previous one by about an inch. Use a utility knife to cut off any extra foil at the end of the tank and also the foil over the tank fittings.

Now construct a tank support frame to hold the tank in place in the collector. If the tank's cold water inlet is located on the bottom of the tank, then the support

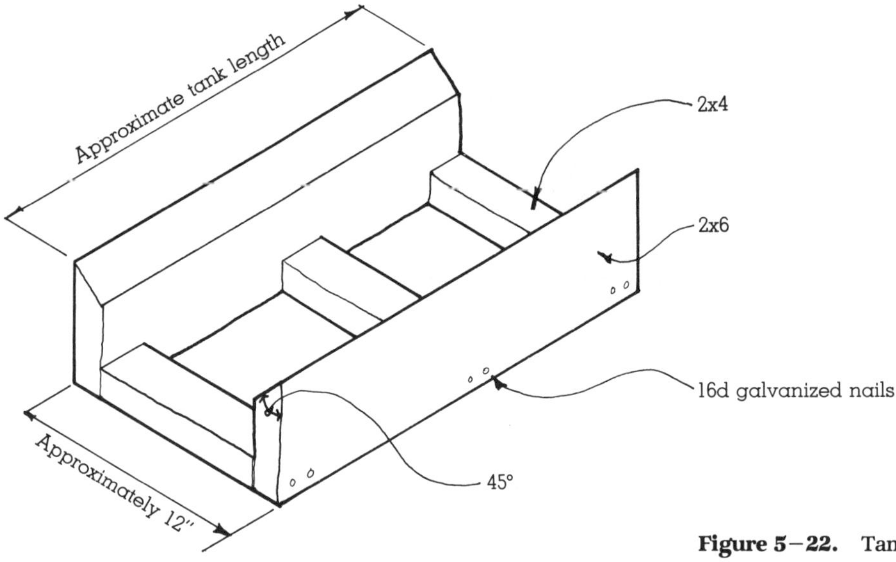

Figure 5–22. Tank support frame.

frame must raise the tank enough so that the inlet pipe can fit beneath the tank and also be pitched slightly downward toward the house. The tank support is made of two 2 x 6s and three 2 x 4s. The 2 x 6s should be approximately equal to the length of the tank and have a 45° angle along one edge. The 2 x 4s are approximately 9'' long and can be cut to slightly different lengths to pitch the tank to one side. All pieces of the support should be painted white.

It is easiest to fasten the tank's fittings before the tank is placed inside the collector, and then measure, cut, and install the last connections with the tank in its permanent resting place. The tank fittings consist of the inlet and outlet water pipes, pressure and temperature relief valve, and vacuum breaker.

First, determine the exact locations and lengths of the pipes and fittings so that they will fit neatly in the collector and be able to connect to the pipes passing through the house wall. You can temporarily place the tank on its support to make the measurements.

The pressure and temperature relief valve should be plumbed so that the end of the relief pipe is directed to the house basement or toward a corner of the collector, possibly to a drain, and away from the glazing.

The inlet and outlet pipe assemblies inside the collector usually consist of a threaded male adapter, several lengths of copper pipe, copper ells, and a union at the end for the connection to the pipes passing through the house wall (3/4'' diameter flexible copper water pipe can be used instead of soldering rigid copper pipes and fittings).

Place the tank on its support frame inside the collector. The tank should be approximately centered between the sidewalls of the collector. Complete the remaining plumbing. When soldering inside the collector, use a piece of metal flashing as a heat shield to protect the collector. Once the plumbing to the backup water heater has been completed, it is time to test your system for leaks.

Open the two valves to fill the solar water heating system.

Caution: Do not allow the water level in your backup water heater to be lowered while the heater is on! The tank heating elements can be damaged in less than one minute if they are on and are not submerged in water. To keep your backup water heater tank full while filling the solar water tank, allow the solar tank to be filled completely before closing the valve to the backup tank.

Note: To test your plumbing for leaks, let your solar water heater operate for 24 hours, then carefully inspect each connection. Plumbers use special equipment to test a system under high pressure and therefore do not have to wait to complete the system.

If the batch is going to be used during cold weather, wrap the pipes with Frost-Tex heat tape and cover them with a minimum R-8 insulation. Protect all foam insulation from exposure to the sun with either aluminum foil duct tape or PVC pipe.

9. Installing the Glazing

Either double or triple layers of glazing can be used effectively with this collector. My glazing system preference with this design is a single layer of tempered glass as an outer glazing, and two layers of Teflon film for the inner glazings. The glass provides an attractive finish, and the Teflon has high solar transmittance, is moderate in cost, long lasting, and lightweight. When my budget can afford it, I use tempered, low-iron, textured glass on the exterior for its added esthetics and performance.

To install the glazing you must construct two wood frames—the *inner glazing stop* and the *Teflon frame*—and secure them to the inside of the collector's south-facing opening. Make the wood frames from quality-grade wood; knots are not acceptable. Do not use pine for these pieces. Also, the wood should be treated with preservative and sealed with two coats of white exterior-grade paint. Let the paint dry for two days before sealing it in the collector. When exposed to high temperatures, paint and pine can emit gases that can condense on the glazing.

Note: Assembling the Teflon glazing frame is time-consuming. Installing a factory-made insulated glass unit—double or triple glazed—is much quicker, but more expensive and heavier. (Triple-glazed units are available with inner and outer sheets of glass and a plastic film suspended between them.)

The frame closest to the water tank is the inner glazing stop. Use a tri-square to draw the outer edge of exactly where the stop will be positioned. With the glazing system shown, set the outer edge of the stop 1 1/2″ in from the outer edge of the collector's south face. When cutting these glazing stops, note on the illustration

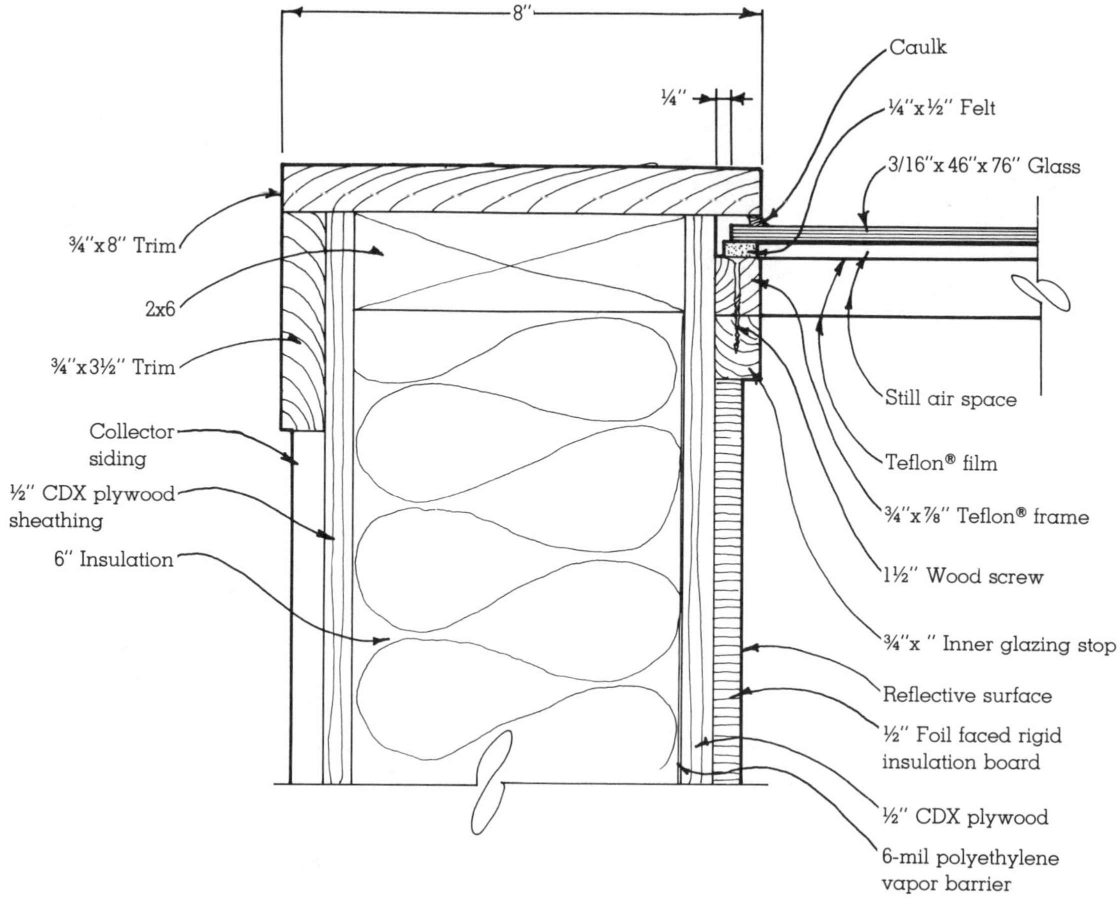

Figure 5–23. Top view of collector sidewall glazing detail.

that the edge of the upper and lower stops, and the ends of the two side stops, must be cut at a 45° angle. Use 4d galvanized finishing nails and exterior-grade glue to secure the glazing stops in place.

After the glazing stops have been mounted you can cut two triangular pieces of 1/2″, foil-faced rigid insulation board for the collector's sidewalls. Apply duct tape on the edges of the foam and secure the foam over the plywood sidewalls. Nails and caulk can be used to mount the foam to the sidewalls.

Construction of the Teflon frame must be done very carefully. The frame must be square to fit properly against the inner glazing stops and to ensure that both layers of Teflon remain tightly stretched. Preparation of the frame should be done

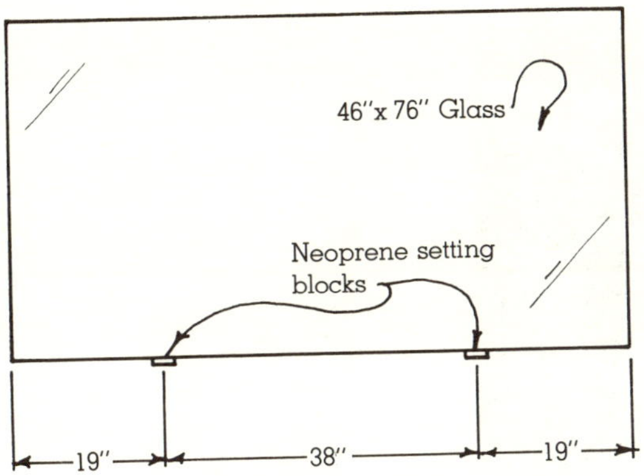

Placement of neoprene setting blocks

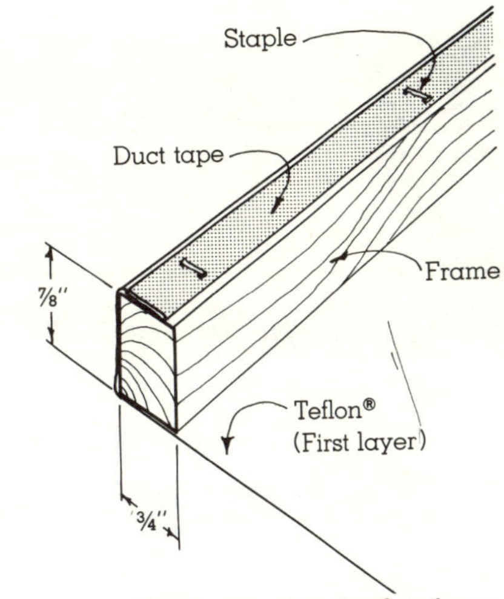

Attachment of Teflon® to frame

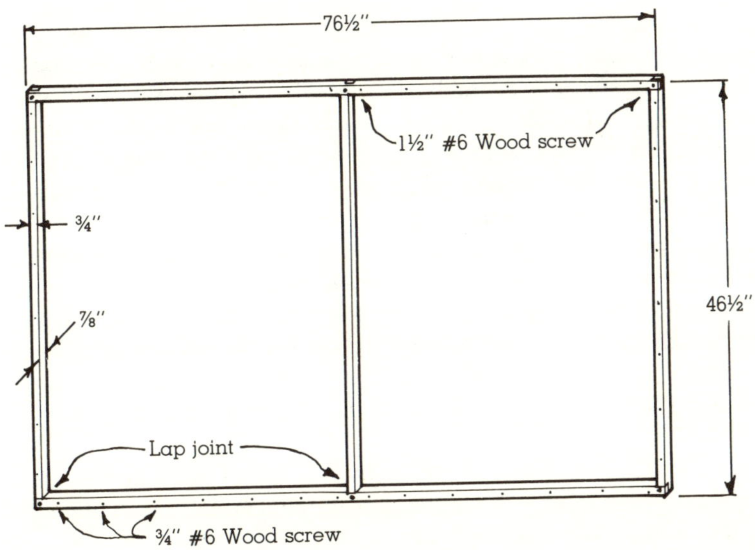

Figure 5–24.
Teflon glazing frame.

on a large (4′ x 7′ minimum), flat table. A sheet of plywood on two sawhorses works fine. Keep the work area clean; any dust will adhere to the Teflon.

The Teflon frame is made of five pieces of 7/8″ x 3/4″ wood—two side pieces, top and bottom pieces, and a center spacing bar. Although the center spacing bar blocks some of the sunlight from entering the collector, it makes the frame more rigid and keeps the two layers of Teflon from sagging and touching each other. All joints should be lapped, glued, and screwed with 1 1/2″, #6 wood screws.

After the frame is painted and thoroughly dried, you can mount the Teflon glazing. A second (and third) person will make this job much easier. Check that

Figure 5–25.
Installing the Teflon glazing.

the frame is square by making certain that the diagonal measurements are equal. Trace the corners of the frame onto the table to use as reference points. Now place the Teflon under the frame. Cut the Teflon with a utility knife or scissors, leaving enough Teflon to wrap around the side and 1/2'' onto the top edge. Refer to figure 5–24. Now wrap 1'' wide duct tape around the edge of the Teflon. Starting at the center of one long side, staple through the duct tape to secure the Teflon to the frame. Work carefully toward the ends and make certain the Teflon is stretched tight. Staple every 2'' along the entire perimeter. Keep checking the reference points to ensure that the frame is square. If you are working alone, you can keep the frame square by placing several finishing nails tightly around the outside of the frame.

When one sheet of Teflon is secured, flip the frame over and repeat the process with the second Teflon glazing.

Use twenty-six 1 1/2'', #6 wood screws or 6d galvanized finishing nails to secure the completed Teflon frame to the inner glazing stop. Be very careful not to damage the Teflon when mounting the frame.

Note: Metal or plastic frames such as those used for storm windows may be a good alternate way to secure the Teflon.

Now staple 1/4'' thick, 1/2'' wide felt weatherstripping on the outer side of the frame. The felt along the bottom of the frame must be placed along the upper edge so that the 1/4'' *neoprene setting blocks* for the glass can fit beneath the felt (see figure 5–24).

Place two 1/4'' x 1/2'' x 2'' *neoprene setting blocks* along the top of the cant strip tucked beneath the felt weatherstripping. Each block should be approximately 18'' from the end of the glass. Clean the glass thoroughly and then lower it into the collector against the felt. The difficult part is now complete.

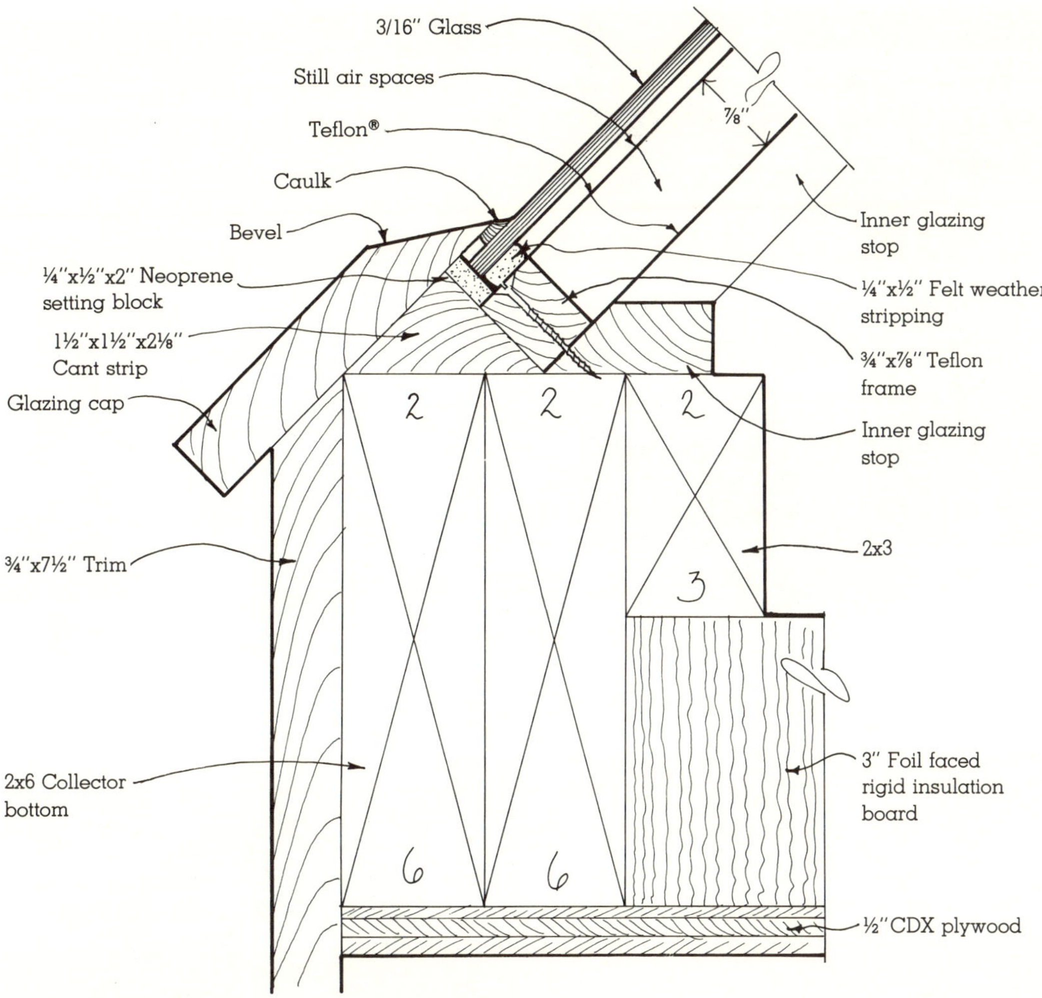

Figure 5–26. Side view of glazing detail at collector bottom.

10. Finishing Touches

To finish the collector you must install the glazing caps, trim, siding, flashing, and caulking. First nail the *glazing caps* in place. Bevel the lower cap to shed water running down the glazing. Now complete the sidewall trim and siding. If you are using wood shakes, work from the 45° angle side toward the house. Slide a piece of flashing under the house siding and over the upper glazing cap. Use silicone or urethane caulk to seal the 1/8'' seam between the glass and the glazing caps, and where the collector adjoins the house.

Figure 5–27.
Nailing the glazing caps in place.

Figure 5–28.
Alternate location for the wall-mounted batch: an attic.

Note: The wall mounted collector design can be modified so that it can be constructed in an attic space with the outer glazing flush with the roof

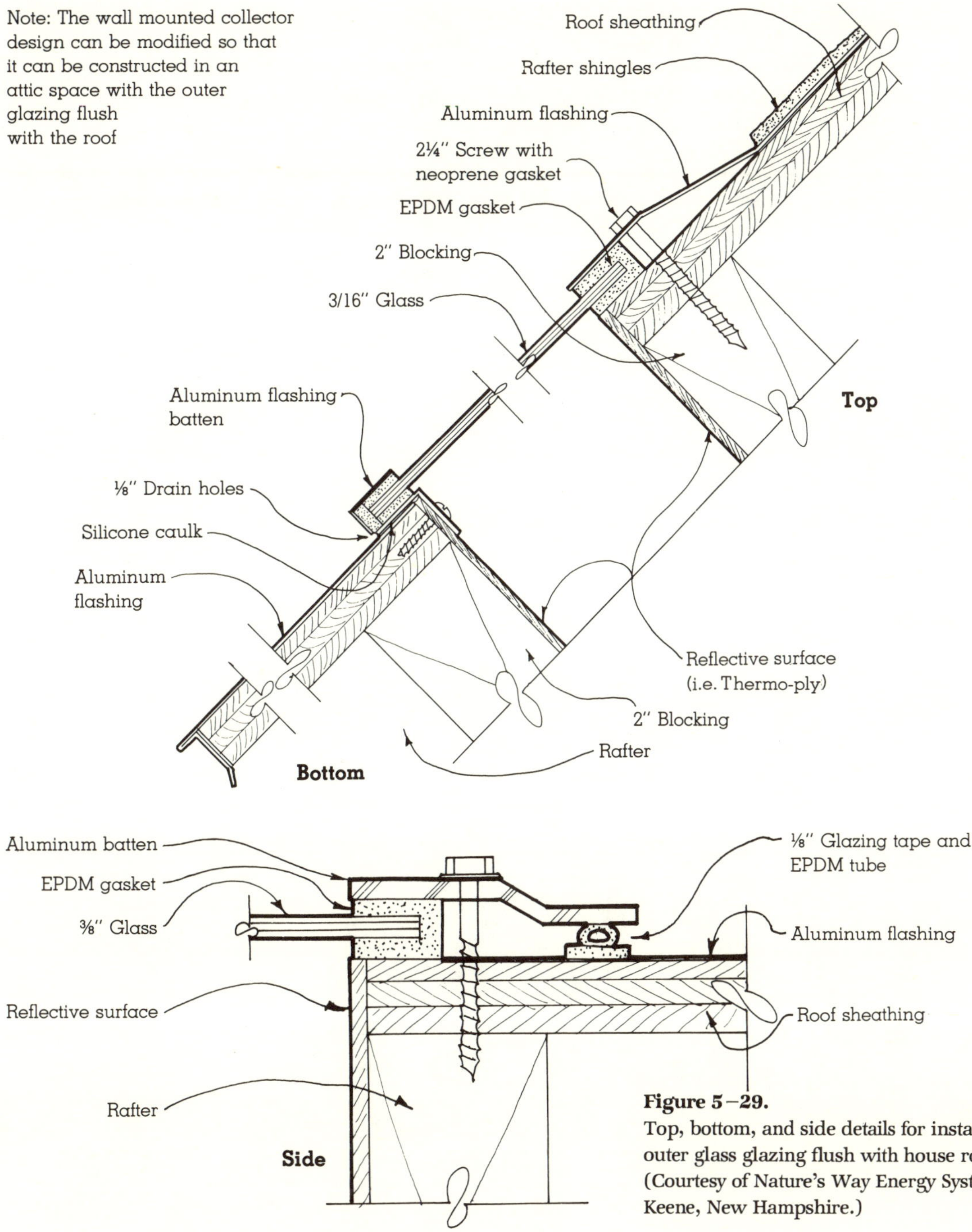

Figure 5–29.
Top, bottom, and side details for installing outer glass glazing flush with house roof. (Courtesy of Nature's Way Energy Systems, Keene, New Hampshire.)

Note: The south-facing glazing caps are exposed to strong sunlight throughout the year. To reduce the harsh effects of the sun on the wood glazing caps, they should be painted a light color. Also, the glazing caps should be made from quality wood. Metal glazing caps are sometimes preferred because of the vulnerability of wood under these conditions.

Option: The wall-mounted collector design can be modified to fit in an attic space with the outer glazing flush with the roof. Figure 5–29 shows the construction details for installing the glazing.

Your batch collector is now finished. Go inside and enjoy your first solar-heated shower!

6 VERTICAL BATCH HEATER

Figure 6–1.
Flat black vertical tank without glazing located in solar greenhouse. (Designed by Irene O'Malley.)

The vertical batch heater is one of my favorite designs—its simplicity is beautiful. This collector is easy to construct, low to moderate in cost, good looking, and can be very effective—as effective as the wall-mounted batch when placed and constructed well. It is ideal for solar retrofits. Because this collector is not structurally attached to the house and, like all batch heaters, is only connected to the house plumbing with two pipes, *this is one of the few solar designs that can be effectively used by renters and those who move frequently.* With a design like this there is little excuse left not to go solar.

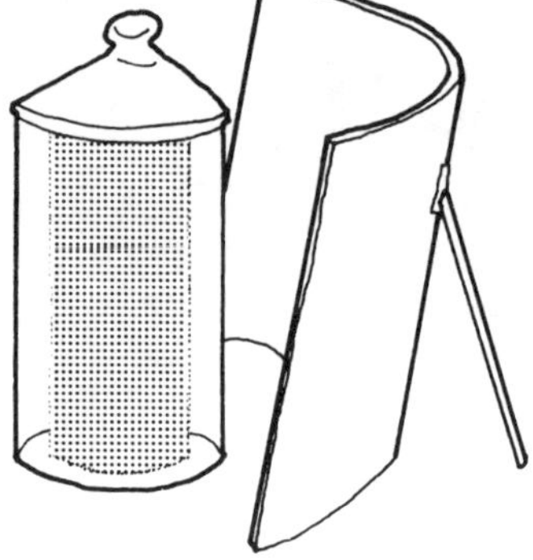

Figure 6–2.
In 1860 Augustin Mouchot constructed a solar cooker and a solar still that looked almost identical to the vertical batch heater. The cooker and still each consisted of a vertical black copper cylinder surrounded by a glass sleeve; a curved mirror was used to reflect additional sunlight onto each tank. The systems were reported to produce excellent beef stew and brandy. (Source: *La Chaleur Solaire*, Augustin Mouchot. Gauthier-Villars, 1879.)

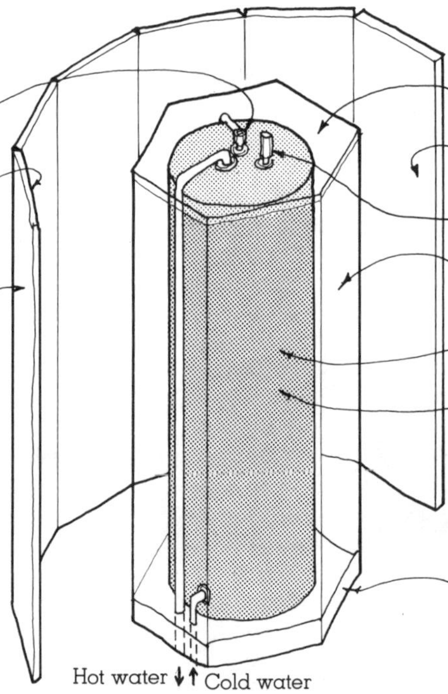

Figure 6–3.
A vertical batch water heater.

DESIGN

This design consists simply of a vertical water tank covered with flat black paint or a selective surface, transmissive insulation on top of the tank, a glass or plastic glazing wrapped around the tank, and a reflector behind and beneath the tank. The reflector is the key that can make the difference between warm and hot solar-heated water. The reflector can even be constructed so that it can be wrapped around the tank's glazing for night insulation. During the day the open reflector can more than double the amount of solar energy striking the water tank. At night the insulating curtain wrapped around the glazing more than doubles the insulating value of the glazing.

I first learned of the concept of a vertical batch from Steve Baer of Zomeworks. His Pro Heater uses double-walled acrylic glazing and a removable, 2″ thick, Styrofoam insulating cover; the collector is located in front of a reflective south wall. Plans for the Pro Heater are available from Zomeworks Corp.; see page 37.

Note: In general, vertically oriented solar collectors perform best when the sun is low in the sky. Therefore, in winter they can perform as well or better than sloped collectors, but in summer the vertical collector is at a disadvantage.

LOCATION

The vertical batch heater can be located in any sunny area outside (not necessarily the south side of the building) or it can be placed inside a south-facing room or greenhouse. It can be placed on a flat roof or, with some modification, on a sloping roof. No matter where you locate this collector, remember that the use of a reflector is a *must* for good performance.

Indoors

When the collector is located indoors, the amount of sunlight striking the collector is reduced by the roof, walls, and structure of the house; additional light is lost through absorption and reflection as the light passes through the house glazing. For good performance, this reduction in the amount of light indoors must be overcome by exposing the tank to *full sun throughout the day* and by using a large reflector. Study the sun angles carefully; without full sun the tank will rarely exceed 100°F. The major advantage of constructing this collector indoors is to protect it from the weather, especially potentially freezing outdoor temperatures. Indoor collectors can be extremely simple and therefore low in cost. In cold climates, having the collector in a 65°F indoor environment, as opposed to the cold outdoors, can sometimes more than compensate for the reduced amount of sunlight indoors. In fact, during months of below-freezing temperatures, the outdoor collector would be shut down completely while the indoor collector keeps on working. In cool and mild climates, placing the tank in a warm indoor environment will probably not compensate for the reduced amount of sunlight. Also, allowing additional sunlight into the house during the warmer months may cause overheating of the house.

Another consideration for the indoor batch is the weight of the collector. An average water tank weighs 100 lb. plus approximately 8 lb. for each gallon of water, which makes a 50-gal. collector weigh 500 lb. Most floors can hold this weight (three people standing close together), but the problem is that the floor may

Figure 6–4.
A vertical batch with Exolite (acrylic) glazing under construction.

sag over a period of time. If you cannot determine if your floor is strong enough, consult a builder or building inspector who can make the evaluation.

The pipes from the house plumbing to the indoor batch should be carefully located to minimize circulation and visual obstruction. On the first floor of a house you can often conceal the pipes along the basement ceiling or through a crawl space and have the pipes penetrate through the floor directly beneath the collector—a very clean solution. Another approach is to conceal the pipes in the ceiling above the collector, then have the pipes enter through the top of the collector. Where you must expose the pipes, keeping the details clean and precise will make a very professional-looking job.

Outdoors

When the collector is located outdoors, extra attention must be given to retaining warm water on cold nights. Fortunately, outdoor collectors have less difficulty getting adequate sunlight.

To maximize heat retention in an outdoor vertical batch, double or triple glazing, transmissive insulation, and a selective surface are recommended. Double glazing, transmissive insulation, and flat black paint can produce good results with less initial cost. Night insulation can be used to increase heat retention, thus reducing the need for a third layer of glazing and selective surface. But, outdoor movable insulation can be troublesome to operate and maintain when subjected to the harshness of the rain, wind, and snow.

When the vertical batch is located outdoors, increasing the amount of light striking the collector by the use of reflectors can make the difference between warm and hot water. It is not necessary to have sophisticated reflectors to achieve good results. The reflectors can be as simple as a white or light-colored wall and ground covering. Since these light-colored surfaces reflect light in a nonfocusing, diffuse pattern, the reflective surface should be at least several times larger than the collector.

The reflective wall behind the collector should be located at a distance between one-half and one times the tank's diameter. Additional light can be reflected onto the tank by curving the reflector or adding another reflective surface to the northeast or northwest of the tank.

The outdoor collector can be located in any sunny area. By installing the inlet and outlet water pipes underground you can place the collector away from the house. If there is little chance of freezing temperatures in your climate or if you drain your collector before below-freezing temperatures occur, you can simply lay the pipes in a shallow trench. Where freezing is a problem, consider using polybutylene pipes. Water can freeze in this type of plastic pipe and not damage the pipe. (Check your local building code before using any plastic pipe.) Be sure to insulate the hot water pipe from the collector.

Since this collector can be attractive, you can place it where it will have convenient access to your house plumbing, possibly in the middle of your garden. And, with the collector in full view it is more likely to get any attention it may need.

A modification of the vertical tank design is to slope the collector at an angle approximately equal to the local latitude angle. The Crystal City batch (page 29) is a good example. The advantage of the sloped design is that good performance can be achieved with use of a smaller reflector than is needed for the vertical tank design. The sloped collector is ideal for mounting on roofs that slope toward solar south. One potential disadvantage is that it is especially difficult to provide movable insulation on sloped tanks. The sloped collector is also somewhat more difficult to construct, especially on a steep roof.

PERFORMANCE

You should test the performance of the vertical batch in its proposed location before completing the plumbing to the collector. Simply set the collector in position and fill it with water. Measure the temperature of the water you put in the tank, then record the water temperature after several hours, and at the end of the day. Take measurements at both the top and bottom of the tank. Testing the collector performance at several locations and with several reflector arrangements gives you a good indication of what you can expect once the collector is completed. When the collector is connected to the house plumbing, the water temperatures will be lowered by the use of hot water, which causes cold water to enter the solar tank; but with reduced water temperatures the efficiency of the collector is actually increased.

With an indoor location, simply having a bare water tank at room temperature can save 10% to 25% of the energy needed to heat the water to 120°F. Because the temperature of the water is raised by the warmth of the house, some additional house heating will be needed in winter, offsetting some of this savings. On the other hand, in summer the cool tank will help cool the house.

By placing the bare, black tank in a sunny spot in the room savings can be increased to 25% to 35%. When the tank is located in a greenhouse the solar-warmed water can be used directly to water plants.

Concentrating the sunlight onto the tank with a reflector and using one or more layers of glazing, transmissive insulation, and other heat retention devices can raise savings up to 35% to 70%!

CONSTRUCTION

Construction of the vertical batch consists of five basic steps: tank preparation, plumbing, glazing, and building the base and reflector.

1. Tank Preparation

Similar to the wall-mounted collector, the tank can be covered with flat black or selective surface paint or selective surface foil. Using a selective surface produces similar results to adding another layer of glazing. See page 105 for how to prepare the tank for painting or how to apply the selective surface foil.

2. Plumbing

Installing the necessary plumbing to the vertical tank is very straightforward. The tank needs an incoming cold-water supply and an outgoing hot-water return. A pressure/temperature relief valve and vacuum breaker can be installed either at the tank inside the glazing, or elsewhere in the plumbing system. Before installing the plumbing, you will have to determine how far the tank will be raised from the floor by its base.

A single-tank vertical batch can provide excellent temperature stratification. To maximize temperature stratification in the tank, the cold water must be dispersed into the bottom of the tank and the hot water must be drawn from the very top of the tank.

The cold water can be dispersed at the bottom by either of two techniques. The cold-water pipe can enter the tank at the lowest tank fitting. This is usually where the drain valve is located on a water heater. The drain valve can be placed at another point in the plumbing as long as the pipes are continuously pitched downward toward the drain. The second technique is to have the cold water enter at a fitting at the top of the tank and pass through a *dip tube* that directs it to the bottom of the tank. Most water heater tanks come with a plastic dip tube already installed. You can see it by looking into the tank. The top of the dip tube usually has a flange that rests on a built-in seat in the cold water inlet of the tank. If you plan to use the dip tube, make certain that you arrange the plumbing so that the cold water enters the correct fitting. If you try to place the dip tube into the hot water fitting, the tube will probably fall into the tank.

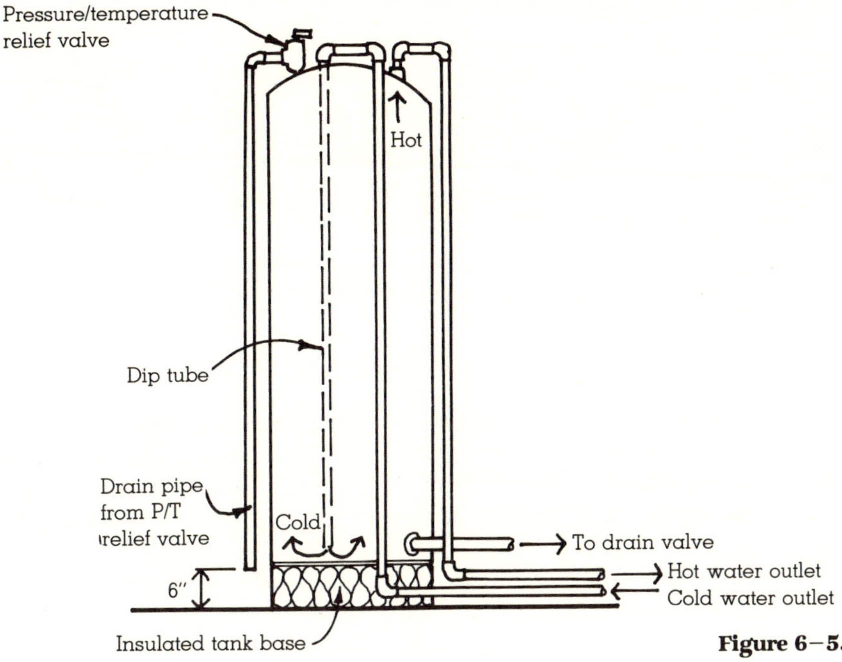

Figure 6–5. Vertical batch plumbing.

The hot water outlet should come from the top of the tank, then down along the side of the tank. Keep the pipe(s) tight against the side of the tank to minimize the size of the glazing enclosure. If you are using a prism shape glazing enclosure, locate the pipe(s) so that it fits into the corner of the prism.

When the tank plumbing is completed you can paint the plumbing with flat black paint.

3. Glazing and Transmissive Insulation

The vertical batch can be enclosed with a wide variety of glazing materials. When double or triple glazing is used, each layer can be selected separately for its best qualities, or all layers can be of the same material. For example, rigid outer glazing such as glass or acrylic can be incorporated with a thin film inner glazing such as Teflon. Flexible plastic sheets (including Sunlite Premium) can be used as inner or outer glazings. The inner glazing must be able to withstand operating temperatures of 150°F. The outer glazing should be strong enough to withstand physical abuse.

Since the glazing is a major portion of the total cost of this collector, using glazing material you already have or can obtain inexpensively can provide a low-cost system. If the collector is located indoors then the simplest of glazings can work.

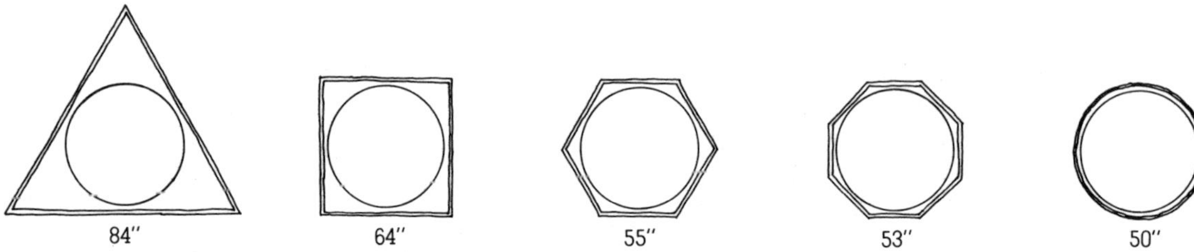

Figure 6–6. The length of glazing needed for various shapes to enclose a 14″ diameter tank allowing a minimum 1″ air space on all sides. (Enclosure top not included.)

The flexibility of the glazing material often determines the best shape for the glazing enclosure. Semirigid plastics, such as fiberglass-reinforced plastics, are easily made into cylinders. If one side of the plastic is coated with an ultraviolet inhibitor, usually the inside on the roll, make certain that the coated side faces the sun.

Glass and more rigid plastics can be made into prisms. Tempered glass is recommended because it is strong and does not shatter. Hexagonal and octagonal prisms are preferable to rectangular prisms because the more sides a prism has, the less surface area is needed to enclose the cylindrical tank. And, with less glazing surface area there is less heat loss. For example, enclosing an 18″ diameter cylindrical area (16″ diameter tank plus 2″ for the pipes and insulating air space) requires a square prism with a perimeter of 72″ or a hexagonal prism with a 62″ perimeter. The hexagonal prism has 14% less surface area and therefore 14% less heat loss than the square prism. Heat gain in both collectors is equal because the tank is the same.

Prisms can be made either with or without a frame. Without a frame, the rigid glazing material is first cut into panels, then sealed together with clear silicone or urethane caulk, or waterproof, high-strength tape. Acrylics can also be bent to shape with a hot wire. With double-walled acrylic panels such as Exolite, you must first cut the inner layer of the panel before bending the outer layer. Making perfect bends with a hot wire is difficult. First practice on some scrap plastic. Two or three other people will make this job easier. When a frame is used it can be constructed of wood or metal. Almost any glazing material can be set within the framed panels. Leave about a 1″ space between the inner glazing and the tank for the necessary plumbing and as an insulating air space.

Thin plastic films can be used as inner glazings. I recommend Teflon because of its durability at high temperatures. A rigid frame can be constructed to support the film. If wood is used to make the frame, do not use pine (spruce or fir are good) and paint the wood with several coats of white, exterior-grade paint. Let the paint dry for at least two days before sealing it into the collector. When exposed to high temperatures paint and pine can both emit gases that can condense on the glazing.

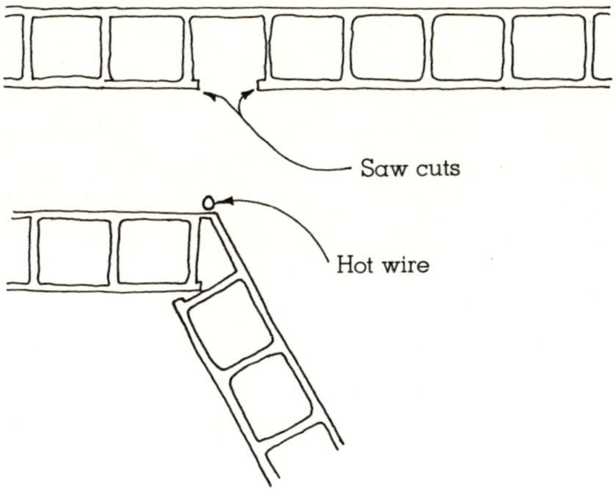

Figure 6–7.
Cutting and bending acrylic glazing.

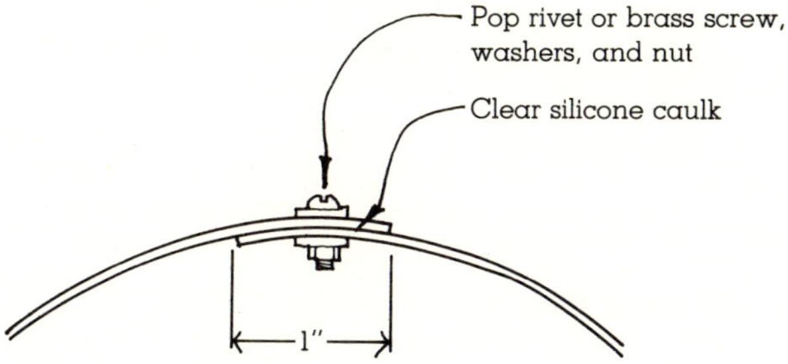

Figure 6–9.
Lap joint for fiberglass-reinforced plastic.

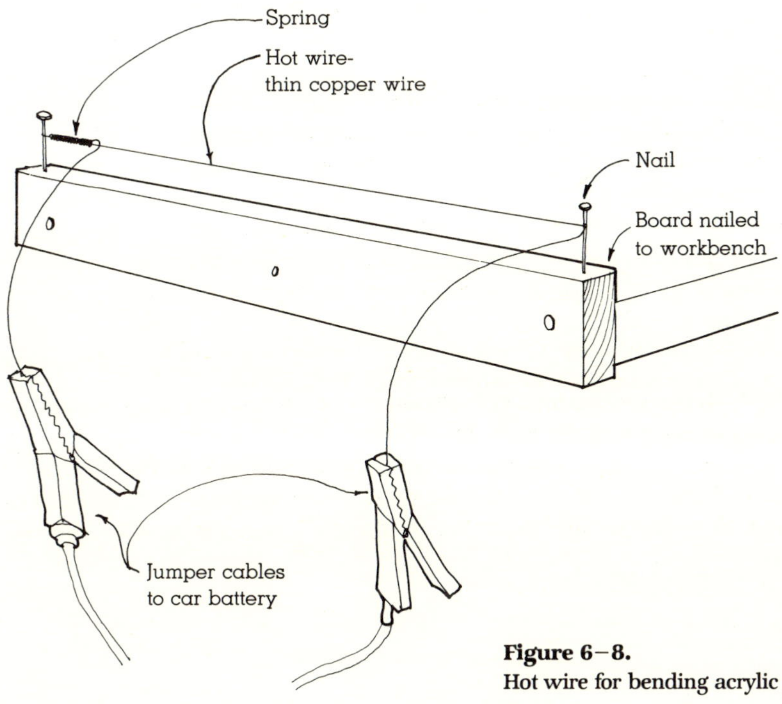

Figure 6–8.
Hot wire for bending acrylic plastics.

Another way to support an inner glazing of film is to use glazing tape to secure the film to the more rigid outer glazing or the tank. Several layers of tape can be built up to increase the thickness of the air space.

The top of the glazing should allow approximately 1'' above the uppermost plumbing fitting, usually the pressure/temperature relief valve. Place several inches of white, semirigid fiberglass furnace filter material in the space above the tank. The fiberglass acts as *transmissive insulation,* allowing much of the light to pass through it to reach the tank and at the same time providing extra insulation for the hottest water at the top of the tank. The low cost ($1 in 1982) of this transmissive insulation is one of the most cost-effective methods of improving solar performance. However, the use of transmissive insulation around all sides of the tank is not recommended because the reduced amount of sunlight outweighs the heat savings from the additional insulation. But, a little transmissive insulation on the north side of the tank might help (See chapter 3.)

4. Wood Base

A wood base can be constructed to secure the glazing around the tank and to raise the tank to insulate underneath it. The base can also raise the glazing above the floor so that the inlet and outlet pipes can enter the collector without penetrating the glazing (a must for tempered glass glazing). Or the outer glazing can be sized to touch the floor, eliminating the need for wood trim. With some rigid glazings (e.g., Exolite) it is difficult to make the glazing into a perfectly symmetrical shape. Therefore you should use the completed glazing enclosure as a pattern to match the base to the glazing.

To construct the base, first trace the shape of the glazing onto 1/2'' plywood or 3/4'' boards. If wood trim is going to be used around the glazing, add 1/8'' around the perimeter of the tracing to provide for tolerance for the glazing fit and for caulk. Next cut the bottom piece of the base, following the traced outline. Use 6d nails to secure the pattern-cut pieces to 2'' x 4'' supports. When using wood trim pieces, miter each corner. For a hexagonal prism each miter cut should be 30°; for an octagonal prism, 22 1/2°. Make certain to stuff the base with insulation before nailing the trim in place.

5. Reflector

For the water in the vertical tank collector to get hot, a reflector must be used to shine additional light onto the tank. Because the reflector is stationary while the sun moves (although some reflectors track the sun), only part of the reflector is useful during different parts of the day and year. Therefore, the reflector must be at least several times larger than the collector to provide maximum solar benefit.

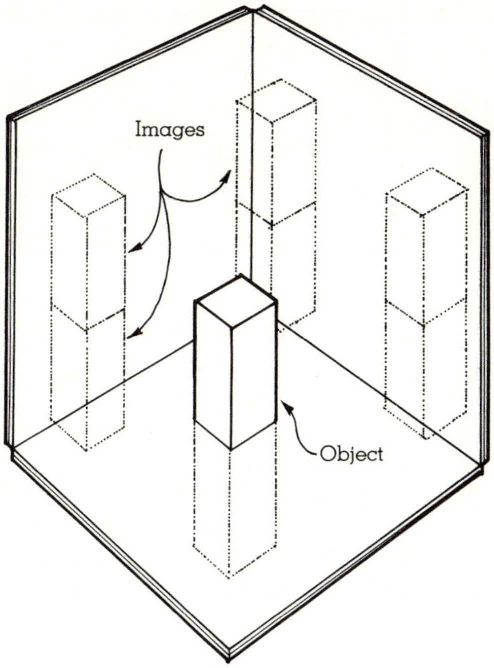

Figure 6–10.
With a corner built of three mirrors, a single object will be reflected seven times—an indication of how reflectors can amplify solar heat. (Source: Pro Heater instructions, Zomeworks, Albuquerque, New Mexico.)

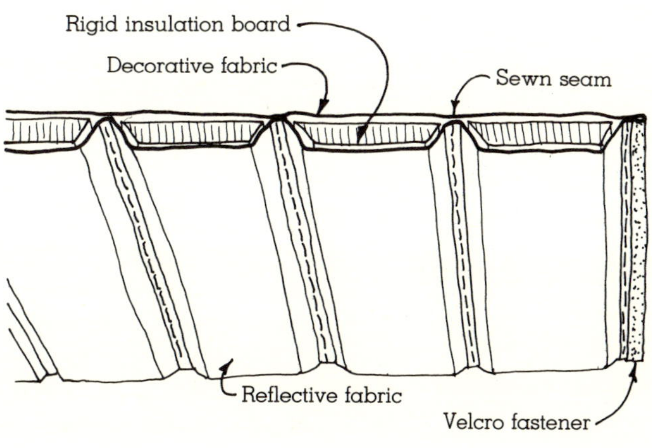

Figure 6–11.
Reflector/insulating curtain.

Steve Baer of Zomeworks, a strong advocate of the use of reflectors, designed this experiment to show the magnifying power of reflectors. When you hold a mirror behind an object, you see the object and its reflection. If your eyes are where the sun would be, the sun would shine once on the object and be reflected once on the object. A second mirror held at a right angle to the first, instead of showing only one more image, doubles the number of objects you, or the sun, would see: one object, two images, and one image of an image. A third mirror placed beneath the object so that the object is now sitting in a reflective corner, doubles the number again: one object, three images, three images of an image, and one image of an image of an image. If the mirrors are large compared to the object, the object and seven reflections remain in view even as you move your head from side to side, simulating the movement of the sun. Using this remarkable concentration of sunlight you can get very hot, even boiling water.

A reflector/insulating curtain can be constructed from two pieces of fabric (one aluminized) and rigid foam insulation board. The foam provides the rigidity so that the curtain can stand upright, and the fabric serves as the hinge, reflector, and decorative covering. Velcro can be used as an operative seal for the curtain. This curtain is designed for indoor use, but with modification can be used outdoors.

For a reflector/insulating curtain that will fit tight around the batch, the inner fabric should be one or two feet longer than the perimeter of the glazing, and the outer fabric should be several inches longer than the perimeter. Both pieces should be at least as tall as the top of the collector. The inner fabric that wraps against the glazing should be reflective. Astrolon or other aluminized materials are suitable and are usually available at large fabric stores. The outer fabric is decorative, but must be able to withstand the wear of daily handling.

The two pieces of fabric are sewn together with full-length vertical seams creating pockets for the rigid insulation to slide into. The inner fabric is slacked to provide for the thickness of the insulation board; 1″ or 2″ thick foam is suitable. (For indoor use R-5 minimum is recommended. Outdoors use at least R-10.) For prism-glazing shapes, each pocket should cover one side of the glazing. For a good fit, miter-cut the foam board edges. With circular glazing any number (more than three) of pockets will work. Velcro can be sewn along the two vertical edges to provide a tight nighttime seal. A separate top can be constructed, but thick transmissive insulation reduces the need for this extra.

If the reflector is not going to be made into an insulating wrap, consider making an even larger reflector.

For an outdoor reflector a light-colored wall and ground area are simplest and can work well because of their size. White gravel is a good ground reflector. If night insulation is desired, then a reflector/insulating curtain can be made with waterproof fabric and fastened each day to stakes anchored in the ground.

Another outdoor reflector/insulating shutter design uses a single pivot point and two curved pivoting shutters. The pivot point can be secured to the ground or wall of the house. With tight night insulation, this collector is safe from freezing in all but the coldest weather.

7 PLUMBING BASICS FOR THE BATCH HEATER

Figure 7–1.
A typical three-valve bypass system. (Note the vacuum breaker on the top of the tee at the right.)

Plumbing the batch is unique for each installation. Although you may have never previously installed plumbing, you will find that with careful planning and good workmanship you can complete the necessary plumbing flawlessly. Always consult your local building inspector to receive approval of your plumbing plans.

Some areas restrict the use of certain materials or only permit plumbing to be installed by a licensed contractor.

Note: Inexperienced home owners who install their own plumbing sometimes accept less-than-perfect work. If there is any leak whatsoever, fix it—a leak wastes water and energy, and can cause further damage.

There are three sections to plumbing your system: connections to the solar water tank, pipes installed between the solar water tank and the backup water heater, and connections to the cold water main and backup water heater. Plumbing is similar to Tinker Toys: you assemble a variety of connectors and pipes to make a path from one place to another. Since pipes and plumbing fittings are available in copper, iron, and several types of plastic, you have to determine which material is best for your needs.

Note: Eventually all water tanks leak. Glass-lined tanks usually last about ten years. Design your system so that when the tank springs a leak, minimal damage will be done to the tank enclosure and the house. Placing a plastic or metal drip pan under the tank works well, as long as you can detect the leak before all 40 gal. have escaped. The tank enclosure should also allow for repair of the leak or replacement of the tank.

SOLAR WATER TANK

Solar water tanks have as many as six fittings. The two most important to the operation of the solar water heater are for the cold water inlet and hot water outlet pipes. The other fittings can be used for a pressure/temperature relief valve, vacuum breaker, anode rod, and drain valve.

To receive the warmest water from the batch, the plumbing system should be designed to encourage temperature stratification, that is, separation of warmer and cooler water. Since the warmest water in the tank automatically rises to the top, the outlet water pipe should be located at the top.

Install the inlet pipe so that the cold water enters at the bottom of the tank. A cold water *dispersion tube* can be used to reduce the mixing of incoming cold water with the hot water already in the solar tank. This tube disperses the water gently into the botton of the tank, rather than allowing the cold water to gush into the tank. Since dispersion tubes are not standard plumbing fittings, you will have to custom make your own.

In single-tank systems where both the inlet and outlet pipe connections are on the same end of the tank, a *dip tube* can be used to separate the incoming cold water from the outgoing hot water. Most tanks come with a plastic dip tube inserted in the cold-water fitting. You can see it if you look into the tank. For horizontally mounted single-tank systems you can make a custom dip tube from either flexible copper or plastic tubing, or soldered copper pipe to further separate the hot and cold water.

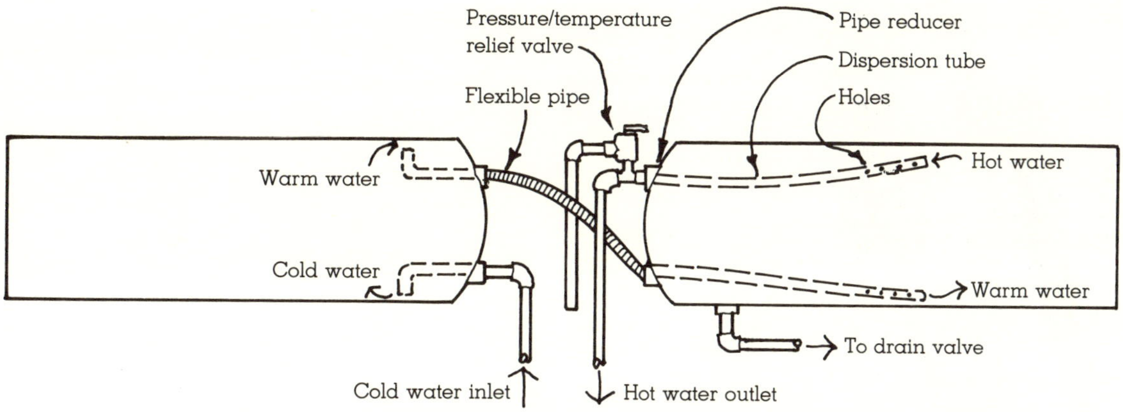

Figure 7–2. Horizontal two-tank plumbing schematic. *Note*: To increase the temperature stratification in each tank, use either ell-shaped assemblies or long, flexible dispersion tubes with holes drilled in the ends.

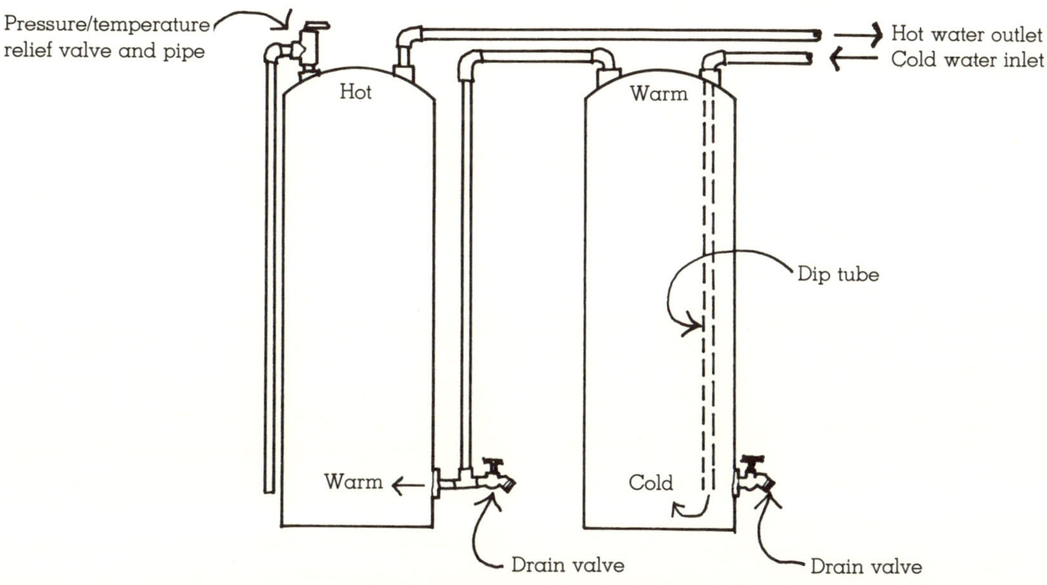

Figure 7–3. Vertical two-tank plumbing schematic. *Note*: To maximize the advantage of temperature stratification, the cold water must be released into the bottom of the tank using either a dip tube or by plumbing the cold inlet pipe into a low fitting, and the hot water must be drawn from the top of the tank.

Figure 7–4.
Corroded anode rods. (Photograph by David Bainbridge.)

With two or more solar tanks, plumb them in series to increase temperature stratification from one tank to the other tank.

The *pressure/temperature relief valve* protects the plumbing against extreme overheating and the resulting danger of a steam explosion. Although it is unlikely that temperatures within the solar water tank will reach boiling (when there is at least 1 1/2 gallons for each square foot of glazing), this valve provides inexpensive insurance. The outlet opening of the pressure/temperature relief valve should be plumbed so that any hot water that is expelled is directed to a safe area. Most plumbing codes require this valve.

A *vacuum breaker* allows air to enter the tank when draining the tank. This aids in quick draining. Some plumbing codes require a vacuum breaker so that water cannot be sucked out of the tank by negative water pressure. Negative water pressure can occur when a large amount of water is drawn from the main water supply, such as during a fire.

Most glass-lined water tanks have a "sacrificial" *anode rod* already in them. If you are installing a used tank, check the condition of the rod and replace it if it is corroded. The anode rod is made of a material (usually magnesium) that corrodes more easily than the steel tank. It should be examined periodically. When there is a crack or chip in the glass lining of the tank, the anode rod breaks down before the steel, thus sacrificing itself to protect the tank.

A *drain valve* is only needed when the tank cannot be completely drained through the cold water inlet pipe. The drain can be used to completely drain the tank for freeze protection and to drain sediment from the bottom of the tank. If your solar water tank has a drain valve but you use the cold water inlet to drain the tank, make certain that water does not stay in the valve and freeze. To avoid this you can simply replace the valve with a plug.

AUXILIARY WATER HEATER

The solar water heater can be connected to the auxiliary water heater in two ways: a *three-valve bypass* connection or a *four-valve plus tempering valve* connection.

The three-valve bypass has two modes of operation. In one mode the solar water heater serves as a preheater for the auxiliary water heater. The solar-heated water always passes through the backup heater before going to your faucet. Except

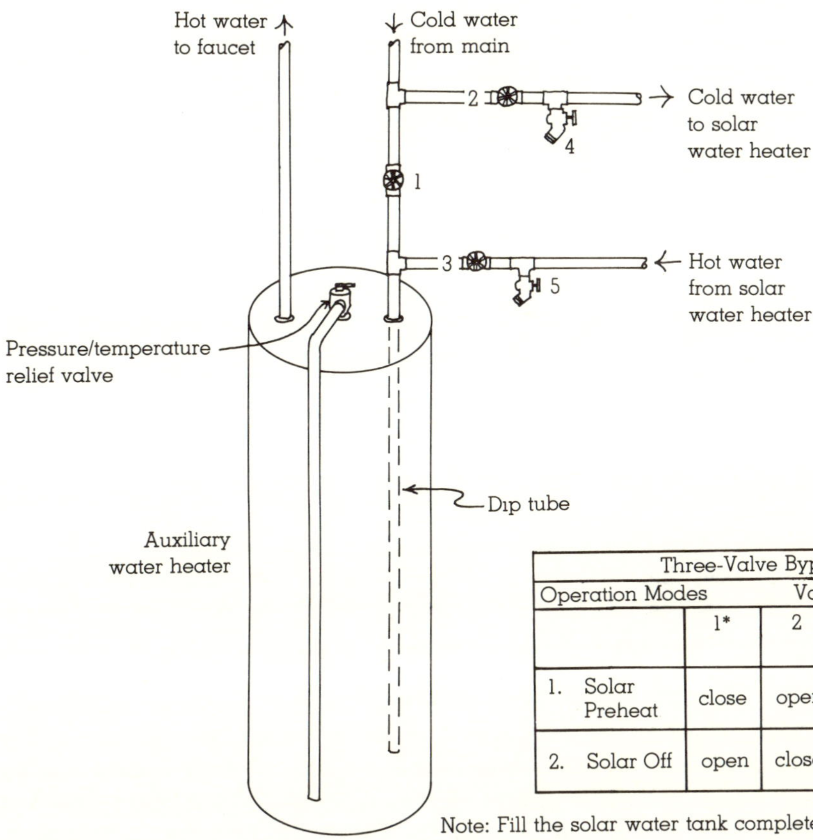

Three-Valve Bypass System				
Operation Modes	Valve Operation			
	1*	2	3	4 5 Drain Valves
1. Solar Preheat	close	open	open	close
2. Solar Off	open	close	close	open

Note: Fill the solar water tank completely before closing valve 1.

Figure 7–5. Three-valve bypass system.

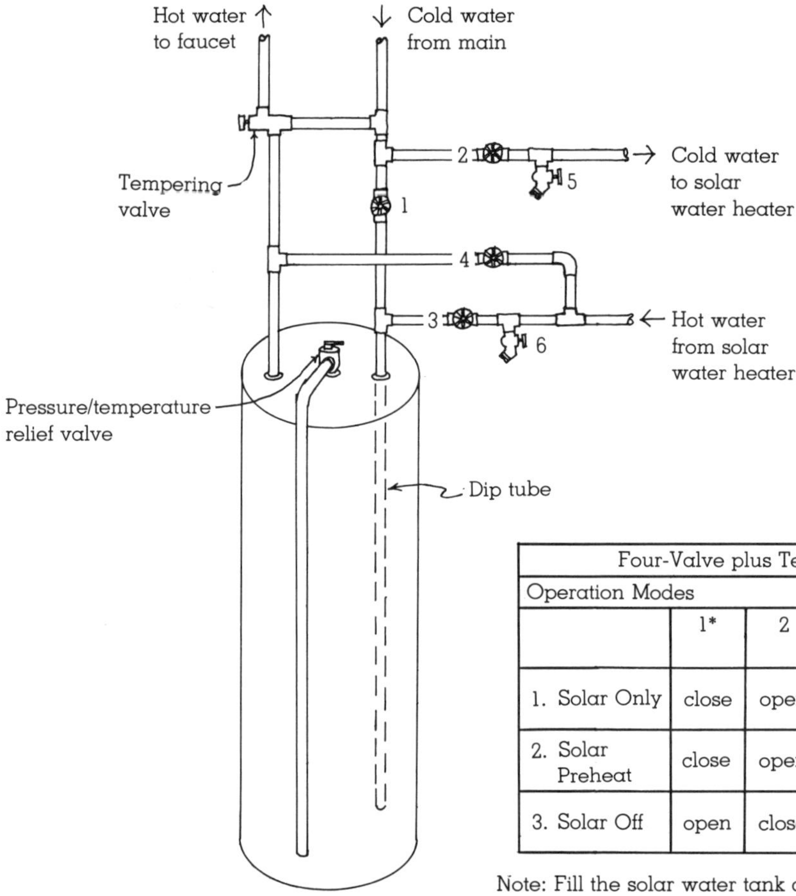

Four-Valve plus Tempering Valve System					
Operation Modes	Valve Operation				
	1*	2	3	4	5 6 Drain Valves
1. Solar Only	close	open	close	open	close
2. Solar Preheat	close	open	open	close	close
3. Solar Off	open	close	close	close	open

Note: Fill the solar water tank completely before closing valve 1.

Figure 7–6. Four-valve bypass system.

for your reduced fuel bill and much greater supply of hot water, you are unaware of the solar performance. With the valves adjusted for the second mode of operation, the solar water heater is bypassed and the conventional water heater does all the work. This allows the solar water tank to be drained.

The four-valve connection allows a third option. You can use the solar-heated water directly and bypass the conventional water heater. The advantage of this option is that you do not have any heat loss from warm water being stored in the auxiliary water tank. The disadvantage of using the solar-heated water directly is that on cloudy or cold days, or during periods of great hot water use, you may run out of hot water.

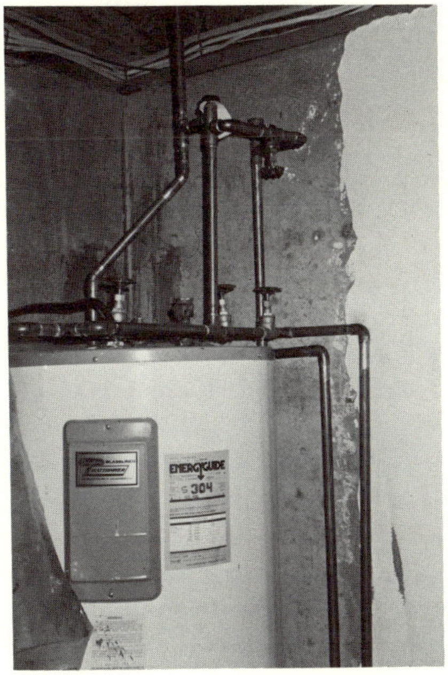

Figure 7–7.
Auxiliary water heater connected to the solar water heater with a four-valve plus tempering valve system.

Caution: If you use the solar-heated water directly, I strongly recommend that you install a thermostatically controlled tempering valve to protect users, especially children, from potentially scalding water.

The auxiliary water heater's hot water pipe should be equipped with a *heat trap.* A heat trap is simply a loop in the pipe that prevents warm water from rising out of the water heater and into the pipes. Without a heat trap warm water continuously thermosiphons through the pipe, making the pipe into a room heater. A manufactured heat trap can be purchased from the Montgomery Ward catalog ($8) or you can easily make one with four ells and some straight pipe.

When you have completed the installation of the valves, clearly label each valve with a marked card wired to it or marking pen, and mount a card nearby with complete operating instructions.

Note: It is important that you understand the control of valves to prevent damage to your auxiliary water heater. Damage can occur if the water level in the auxiliary tank is lowered while the heating elements are on. Therefore, always turn the heating elements off while working on the auxiliary water heater.

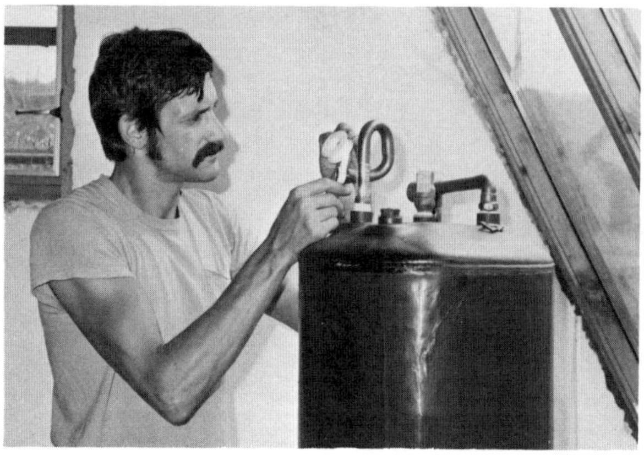

Figure 7–8.
Wrapping Teflon tape around the threaded end of a heat trap.

PIPES

Installing the necessary plumbing between the solar and auxiliary water tanks varies greatly with each installation. Your system should be designed so that the pipe lengths are as short and direct as possible. In cool and cold climates try to avoid having the pipes pass through unheated spaces. To minimize subjecting pipes to outdoor temperatures, locate them inside interior, rather than exterior, walls.

There is a way to install plumbing in an existing wall without having to tear the wall completely apart. First, draw a pencil line along the top of the baseboard molding on the walls where the plumbing will be installed. Next, carefully remove the molding. Access to the wall is achieved by removing the bottom several inches of the wall finish below the pencil line. There may not be ample room to work below the pencil line marking the top of the baseboard molding. In this case you can replace the existing baseboard with 6″ or 8″ wide molding. Now carefully drill through the wall studs; watch out for those nails coming through the bottom wall plate. After completing the plumbing and testing for leaks, install pipe insulation and replace the molding.

Install all pipes so that they are pitched to a low point. A 1/4″ pitch for every foot of pipe is good. Place a drain valve at the low point of each pipe. Two drain valves are usually adequate for draining the entire system.

Before assembling the last pipe lengths, double-check to make certain which is the cold water supply and which is the hot water return. A mistake here would dampen the collector performance and you may never know it.

When the plumbing is complete, insulate the pipes. Insulation of the cold water pipe eliminates condensation on the pipe, which can cause water damage in finished areas of the house. Insulation of the hot water pipe retains some of the solar heat.

Pipe Insulation

Type	*Typical Wall Thickness*	*R-Value Per Inch*	*Cost*	*Installation*	*Temperature Range (°F)*	*Notes*
Elastomeric foam (black) brand: Armaflex	3/4″	4	moderate	easy-pliable around corners	-40 to 220	Breaks down from ultraviolet exposure. Must be completely covered for outdoor use. Shrinkage is a problem, especially if the material has been slit.
Urethane (rigid)	1″	7.2	high	moderate	-40 to 220	Breaks down from UV exposure. Must be completely covered for outdoor use. Cover with PVC pipe or metal jacket. Gives off noxious gases when burning.
Fiberglass (foil or paper-faced batts)	3 1/2″	3	low	difficult; wear gloves and breathing mask; looks rough	-60 to 450	Cut pieces from standard building insulation and tape seams. For indoor use only.
(rigid)	1″	3	moderate to high	moderate	-60 to 450	Must be protected from water and extreme moisture; use PVC pipe or metal jacket for outdoor use.
Phenolic foam (rigid)	1″	approximately 5; varies with manufacturer	high	moderate	-40 to 300	Does not give off noxious gases when burning. Less durable than urethane.
Polyethelene (semi-rigid tubing) brand: Snap-It-On	1/8″	very low	low	easy	-	Insulates by trapping air around the pipe. Not recommended because of low R-value.
(foam)	3/8″	approximately 4	moderate	moderate	-110 to 200	Must be completely covered for outdoor use.

HOW TO SWEAT SOLDER COPPER PLUMBING

Most homes use copper plumbing for water. Using copper plumbing for your batch is generally recommended because it is compatible with your home's plumbing, durable, and easy to install once you learn how.

Sweat soldering consists of seven basic steps: planning, measuring and cutting, cleaning, fluxing and assembling, soldering, cooling, and testing for leaks. First you must plan the plumbing layout, then measure and cut the pipe lengths. All pipes and fittings must be thoroughly cleaned to remove all dirt, oil, and oxidation. Flux is then applied to the cleaned, bare metal to prevent reoxidation and to help the solder flow. Then you assemble the pieces and heat each joint until it is hot enough to melt the solder. The molten solder fills and seals the joint. Next, you wipe the joint with a damp rag to cool the solder and give you a clean-looking, watertight seal. Finally, you must test for leaks and resolder any leaking joints you find.

In order to sweat solder you need the following tools. They are a good investment; once you learn how to sweat solder there are many household jobs you can do. These tools are available at plumbing supply and good hardware stores.

1. propane gas tank
2. torch for propane tank
3. emery cloth
4. plain steel wool
5. 3/4'' fitting cleaner brush
6. 3/4'' pipe cleaner brush (optional)
7. 50% tin/50% lead solder
8. flux and flux brush
9. tubing cutter
10. spark lighter
11. heat shield (a piece of metal flashing)
12. fire extinguisher
13. bucket of water and rag
14. wrench

Copper pipe is available in both flexible tubing and rigid pipe. Flexible tubing (soft tempered) is especially good for tight spots, but you have to be careful not to kink it. Rolls of flexible tubing are available in lengths up to 100′, so that you can have long pipe runs with few connections. Flexible tubing is available in two thicknesses: types K and L. Type L is thinner, but preferred in home use because it is less expensive.

Rigid pipe (hard tempered) is less expensive than flexible tubing and is available in 10′ and 20′ lengths. There are three rigid pipe thicknesses available: types K, L, and M. Type K is the thickest, and most expensive, and rarely used in

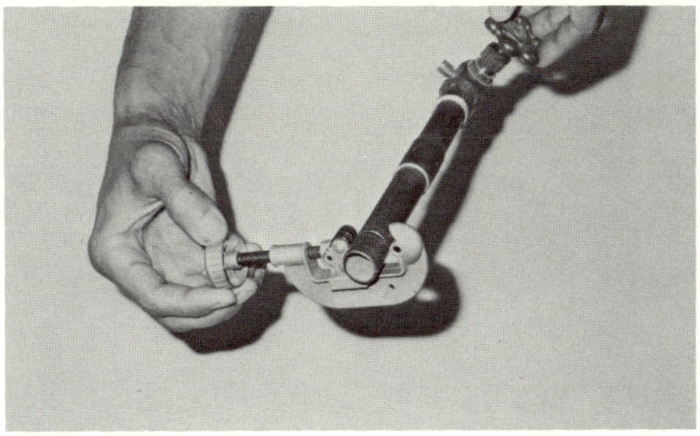

Figure 7–9.
Cutting pipe.

houses. L is the medium weight, and M the thinnest. Type M is most common in residential use because of its lower cost.

Note: When joining certain dissimilar metals, such as copper and steel pipe, you must use special fittings to prevent corrosive electrolytic action between them. Brass fittings can be joined directly to copper and steel making brass a good transition material. Plastic pipe can be joined to any metal pipe with the use of special threaded fittings.

All pipes must be well supported and allow for expansion and contraction. Rigid copper pipe should be supported every 6′ to 8′ and flexible pipe every foot (horizontally). Pipe hangers must be the same metal as the pipe. Plastic pipe can span 2′ or 3′ between supports.

1. Planning

Carefully plan the plumbing layout. Good planning can save you much work and aggravation later. Consider the order in which you will solder the pieces. Ideally you can do most of the soldering in safe, open areas, then slide the assembled sections into place. Plan so that the last few soldered connections can be done where you have ample space to work. Avoid soldering next to wiring, flammable insulation (including foams and some types of cellulose), and gas and oil pipes. When soldering near wood framing, use a piece of metal flashing as a heat shield.

2. Measuring and Cutting Pipe

Always measure the length of pipe needed from the inside lip of each fitting. If the end of the pipe is dented, cut it off with a tubing cutter before measuring. To cut the pipe, simply adjust the tubing cutter's opening so that the cutting blade lightly presses against the pipe. Make certain the pipe is located between the two rollers on the opposite side of the cutting blade. Now swing the cutter completely around the pipe. Adjust the cutter slightly by increasing the pressure of the blade on the

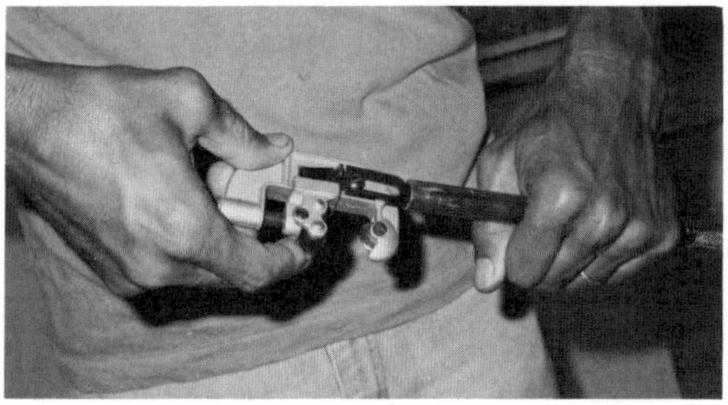

Figure 7–10. Removing burrs inside the pipe.

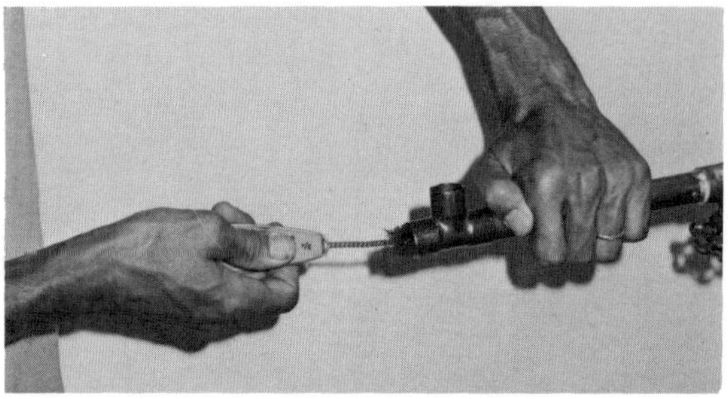

Figure 7–11. Cleaning the fitting.

pipe, and again swing the cutter around the pipe. Repeat this several times until the pipe is completely cut.

After cutting the pipe to the correct length, remove any burrs inside the pipe by pressing the small-pointed blade on the side of the tubing cutter against the inside of the pipe while slowly turning the pipe.

3. Cleaning

All pipe and fitting surfaces that are to be soldered must be very thoroughly cleaned. Any spot you miss now will make *ten times* the work for you later. Also, all pieces must be dry. Wipe dry any existing plumbing that you drain.

Clean the outside of the piece that slides into the fitting (usually the pipe), and the inside of the piece that slides over it (usually the fitting). There are two ways to clean. One way is to first use emery cloth to remove dirt and oxidation, then use steel wool to polish the copper to a bright finish. After every few joints, discard the used emery cloth and steel wool, and continue with clean pieces.

Another way to clean is to use a fitting cleaner wire brush for inside surfaces and a pipe cleaner brush for outer surfaces. Cleaning brushes are available in a variety of standard pipe sizes. You need 3/4″ diameter brushes for 3/4″ pipe. Simply rotate the brush several times around the pipe to produce a shiny finish. The entire surface to be soldered must be cleaned. For short lengths of pipe, the pipe cleaner brush is difficult to handle, so use emery cloth and steel wool.

Note: Valves with rubber or plastic seals should be disassembled before soldering. The washers can easily be damaged from the heat. Use two wrenches to unscrew the lower nut and remove the valve stem.

4. Fluxing and Assembly

Flux must be applied to both the inner and outer surfaces to be soldered. Always apply flux with a brush to avoid contact with your skin. Flux should be applied as a thin, even coat over the entire surface to be soldered.

After fluxing, slide the pieces together. Be careful to avoid getting any dirt on the flux. You are now ready to solder.

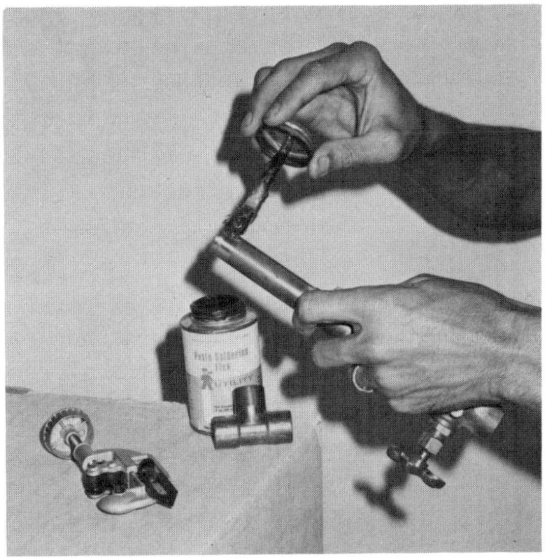

Figure 7–12. Applying flux.

Figure 7–13. Soldering.

5. Soldering

First, place the pipes and fittings in a secure position so that they will not move while you are heating them. If an assistant holds the assembly, make certain there is no movement during soldering and until the solder is cooled. If it is moved before the solder cools, the joint can be weakened.

Set the metal heat shield in place to protect wood within one foot in front of the propane flame. Place the water bucket and rag within arm's reach. You will need them to clean and cool the joint afterward. It is also a good idea to have a fire extinguisher nearby.

Now screw the propane torch head snugly onto the tank. Slowly open the torch valve, then use a spark lighter to light the torch. If the flame pops and goes out, the valve is open too much. Adjust the valve so that the inner bright blue flame is 1'' to 1 1/2'' long.

Hold the torch so that the point of the bright blue inner flame is directly on the fitting, not on the pipe. Do not use the torch to melt the solder. Heat the joint until it is hot enough to melt the solder on its own.

Shortly after the flux begins to flow, the joint should be hot enough. Remove the flame from the joint and touch the solder to the seam of the joint on the *opposite side* from where you held the torch. If the joint is hot enough, the solder will melt and be drawn into the fitting. Continue feeding solder into the joint until the entire joint is filled and a bead of molten solder forms at the bottom of the joint.

If the solder does not melt, remove the solder and resume heating the joint. Then, after another minute of heating, try applying the solder again.

Note: When soldering a separate assembly, hold it so that the molten solder does not flow into an open end. This is especially important on valves and other fittings with threaded ends. Solder that hardens on an unassembled end will have to be cleaned later.

Also, on threaded fittings, first complete the soldering end of the fitting before screwing the threaded end into place. Otherwise, the heat of the soldering can destroy the Teflon tape seal.

6. Cool

Wait approximately 30 seconds after soldering the connection, then take a damp rag and wipe around the joint. The rag cools the soldered assembly and makes a quality, finished-looking joint.

7. Testing for Leaks

After completing the plumbing you should test for leaks. Reassemble any valves you disassembled for soldering. Now adjust the necessary valves to reroute the water flow to the solar tank.

Where possible you can adjust the valves to drain off the water that first passes through the new plumbing. The inside of the connections have small amounts of residue in them that discolor the first few gallons of water passing through them. If you do not drain this water, then you can simply run the faucets for a few minutes until the residue passes through.

Now, slowly open a hot water faucet. Water and air will spurt out for a minute or so until all of the air is removed from the new plumbing. After the air is completely removed, close the faucet and go back to the new plumbing to check for leaks. Look carefully at all joints and valve fittings. If a valve fitting is leaking, tighten it only enough to stop the leak; leave some space in case you have to tighten it more in the future. If a soldered joint is leaking you will have to drain that section of plumbing, dry it, and resolder it.

After your system has been operating for a day or more, carefully inspect for leaks again. Remember that even very small leaks do not stop by themselves and must be repaired.

8 MEASURING BATCH HEATER PERFORMANCE

Figure 8–1.
An effective batch water heater located in a solar greenhouse. Its performance can be measured to determine fuel savings and batch efficiency under various conditions. (Source: Nature's Way Energy Systems.)

There are two basic ways to determine the performance of your batch: comparison and calculation. Comparison gives you the "bottom line" of what you are really saving. Using your fuel cost, you can determine how much money you are saving per month and per year. Calculations provide you with the amount of heat your batch gains or loses in a given situation. You can determine how much heat is gained on a sunny or cloudy hour or day, or how much heat is lost overnight.

The easiest way to determine collector performance is to compare fuel consumption before and after the batch was installed. Make certain to compare the amount of fuel used, and not the ever-increasing cost. To determine the amount of money saved multiply the fuel savings by the fuel cost. Since the amount of hot water used and the temperature of the incoming ground water may vary from season to season, compare fuel consumption for the same month for each year. If the fuel used to heat your water is not separated from other uses on your bill, you can estimate your savings if you can assume that your other fuel uses remained constant from year to year.

CALCULATING SOLAR HEAT COLLECTION

To calculate the amount of solar heat collected by the batch you must first determine several factors: the temperature of the cold water entering the tank and of the solar-heated water in the tank, and the heat storage capacity of the water in the solar tank.

The cold water inlet temperature can be determined by running cold tap water on a thermometer for several minutes. This is the same cold water that enters your collector whenever hot water is used.

To determine the average temperature of the water in the collector use the drain valve on the hot side of the collector. First drain the water that has been sitting in the pipe. Then measure the temperature of a bucket of water removed from the collector. You will actually be recording the temperature of the hottest water at the top of the solar water tank. The average water temperature depends on the amount of temperature stratification within the solar tank. Temperature stratification varies greatly with the system design (one or two, vertical or horizontal tanks), and with the amount of solar gain. In some systems water can also be drawn from the bottom of the solar water tank to provide accurate stratification data. (The average water temperature is approximately halfway between the upper and lower tank temperatures.) Otherwise, you will have to guesstimate the average water temperature after measuring the outlet water temperature.

To determine the heat storage capacity of the water in the solar tank you will have to calculate how many cubic feet of water are in the tank and then multiply the number of cubic feet by the heat capacity of water. Since there are 7.5 gal. in 1 cu. ft., divide the number of gallons in your collector by 7.5. The heat capacity of a material is defined as the number of Btus required to raise 1 cu. ft. of material, 1°F. The heat capacity of water is 62.4 Btu/cu. ft. × °F. It takes 62.4 Btu to raise the temperature of 1 cu. ft. of water 1°F. The heat storage capacity in Btus of the water in the collector equals the number of gallons of water divided by 7.5 and multiplied by 62.4.

To calculate the amount of solar heat gained by the water in the collector, multiply the difference in temperature between the cold water inlet and solar tank (△T, delta T) by the heat storage capacity of the water in the collector:

$$\text{solar heat gain} = \Delta T \times \left[\begin{array}{l}\text{cu. ft.}\\ \text{of water}\end{array} \times \frac{62.4\ \text{Btu}}{\text{cu. ft.} \times {}^\circ\text{F}} \right]$$

For example, if the cold water inlet temperature is 50°F, and the water temperature in the solar tank is 120°F (the hot water outlet temperature was 130°F, but I assume the average water temperature in my one horizontal tank system was 120°F), and there are 50 gal. of water, then the solar heat gain equals:

$$70°\text{F} \times \left[50 \text{ gal.} \times \frac{1 \text{ cu. ft.}}{7.5 \text{ gal.}} \right] \times \frac{62.4 \text{ Btu}}{\text{cu. ft.} \times °\text{F}} = 29{,}100 \text{ Btu}$$

The following equivalents below can be used to interchange Btus with fuel savings.

Oil (@ 65% efficiency) = 93,600 Btu/gal.
Natural gas (@ 75% efficiency) = 75,000 Btu/100 cu. ft.
LP gas (@ 75% efficiency) = 70,000 Btu/gal.
Electricity (@ 100% efficiency) = 3,412 Btu/kwh
Wood (hardwood @ 50% efficiency) = 12,000,000 Btu/cord
(softwood @ 50% efficiency) = 7,000 Btu/cord
Coal (@ 60% efficiency) = 7,500 Btu/lb.

For example, the 29,100 Btu saved equals the amount of energy from 8.5 kwh of electricity.

$$\frac{29{,}100 \text{ Btu}}{3{,}412 \text{ Btu/kwh}} = 8.5 \text{ kwh}$$

EFFICIENCY

To calculate the efficiency of the collector, divide the number of Btus collected by the number of Btus striking the collector. The number of Btus collected was determined in the preceding example by multiplying the difference in temperature between the cold water inlet and the solar tank times the heat storage capacity of the water in the collector. The number of Btus striking the collector, called insolation (not to be confused with insulation), depends on the collector's size, slope, and orientation away from solar south, the amount of shading on the collector, and the amount of available sunshine.

After completing a solar site analysis, you should know both the collector's orientation away from solar south and the approximate amount of shading. The available insolation (sometimes called "solation") varies according to the weather and the slope of the collector surface.

The use of reflectors is a significant variable in calculating collector efficiency. One approach is to consider only the light reflected onto the solar water tank and not the total sunlight striking the reflector. If 10,000 Btu strike the reflector, but only 5,000 Btu are reflected to the tank, add only the 5,000 Btu to the amount of light reaching the collector. A major limitation of calculating efficiency is that by using a reflector as simple as the light-colored south wall of the house, you can

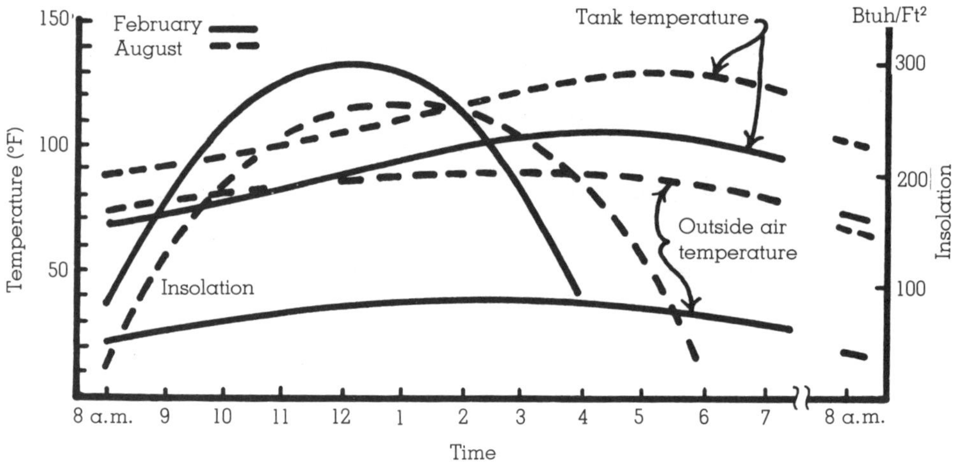

Figure 8–2. Performance of a double-glazed collector with selective surface tank treatment. The collector's efficiency was 50% in February and 55% in August. (Source: Sol-Airesearch, Inc.)

significantly increase performance at no additional cost, but may decrease efficiency at the same time. The hot water produced and the system cost are a much better way to compare water heating systems.

$$\text{collector efficiency} = \frac{\text{solar heat gain}}{\text{solar heat available}}$$

solar heat available = clear day insolation for collector slope (see appendix 9)* × % of maximum solar gain for collector orientation** × [100% available sunlight − % shading on collector] × collector area (including effective reflector area)

For example, a single 50-gal. horizontal tank collects 29,100 Btu on a sunny day in March. The latitude of the site is 32° north. The collector has 25 sq. ft. of glazing sloped at a 45° angle. The reflectors are inside the collector behind the glazing; therefore, their effective area is included in the 25 sq. ft. of glazing area. The collector is oriented 15° east of solar south and has only 5% of the available sunlight shaded from the glazing. What is the collector's efficiency under these conditions?

$$\text{collector efficiency} = \frac{29{,}100 \text{ Btu}}{2{,}358 \text{ Btu/sq. ft.} \times 97\% \times 95\% \times 25 \text{ sq. ft.}}$$

$$\text{collector efficiency} = 54\%$$

*On cloudy days the amount of insolation can be 10% to 90% less.

**If the collector is oriented within 20° of solar south it can receive 95% to 100% of maximum solar gain; for other orientation see page 170.

PERFORMANCE PARAMETERS

Bruce Maeda and Bruce Melzer of Earth Integral in Davis, California, have computer-simulated the performance of a batch collector. Their simulations were based on a collector design by John Burton of Integral Design (figure 8–3). The collector consisted of two 40-gallon tanks in a simple triangular box double-glazed with fiberglass-reinforced plastic. The results of their study were later compared with monitored data of a collector in Davis, California.

The weather in Davis is typical of California's Great Central Valley and of the low desert regions of the southwestern United States. The simulations were used to determine the effect on the percentage of solar contribution of several factors: time of hot water use, auxiliary water heater thermostat setting, overall hot water use, and the use of insulating shutters. Results of the simulations are presented below.

Performance Table

	Water Heater Thermostat Set Point Temperature (MBtu)	
	140°F	110°F
Base case	2.21 20% solar	2.92 45% solar
Standard load profile	1.89 17%	2.57 39%
75 gallon/day, demand	3.63 22%	4.38 45%
Shuttered, 5 pm to 8 am	2.80 26%	3.46 52%

The base case is defined as the afternoon/evening "shifted" load profile, 50 gallons of hot water demand per day and no movable insulation.

Source: B. Maeda and B. Melzer, "A Comeback for Breadboxes," Solar Age, October 1980. Reprinted with permission.

Changing the auxiliary heater thermostat setting from 140°F to 110°F has greater impact on energy savings than either night insulation or shifting the time of hot water demand. For example, with standard water use and a 140°F thermostat setting, only 17% solar was achieved. With the same use of hot water and thermostat setting, but with night insulation the percentage of solar increased to 26%. Shifting the hot water demand toward the afternoon resulted in 20% solar water heating. But with the standard load, changing the thermostat back to 110°F achieved 39% solar use. While greater demand for hot water may not increase the percentage of solar, it does increase the cost-effectiveness of the collector because energy output increases.

Actual experience in Sacramento and Davis shows that this collector design can provide 100% of the hot water needs for two or three adults for six months of the year. During the rest of the year the system acts as a preheater providing approximately 50% of the hot water needs. In these actual cases the hot water demand was reduced with the use of low-flow fixtures, and the demand pattern was changed to one that resembles the "shifted" demand in figure 8-4.

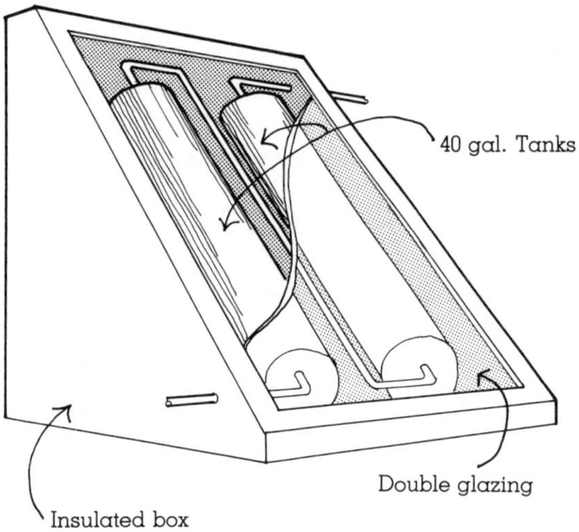

Figure 8–3.
Test collector used in California. (Designed by John Burton.)

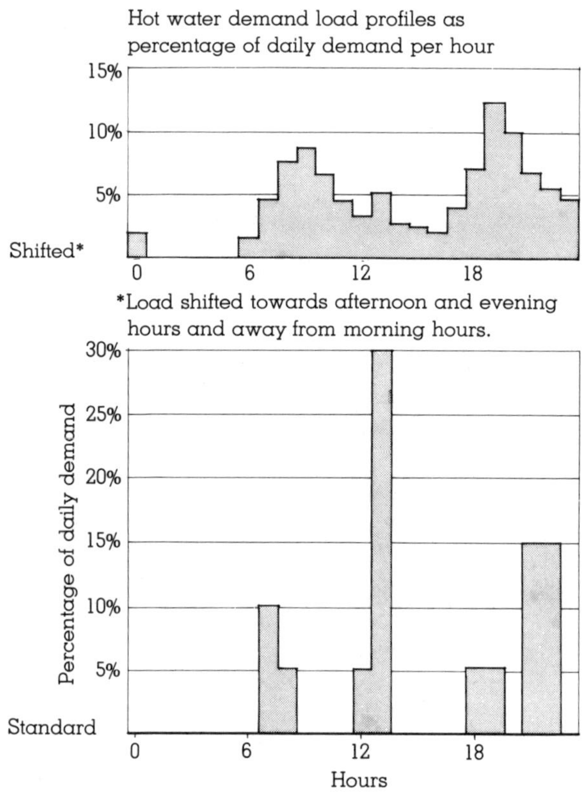

Figure 8–4.
Hot Water Demand Load Profiles as Percentage of Daily Demand per Hour.
(Source: Bruce Maeda and Bruce Melzer, "A Comeback for Breadboxes," *Solar Age* magazine, October 1980. Reprinted with permission.)

9 CONSERVATION: TUNING YOUR WATER SYSTEM

Just as a tight, well-insulated house gives a real boost to solar-heated houses, conservation can make a dramatic impact on solar water heating performance. In most houses *simple conservation measures alone can save 20% to 40% of your water-heating costs!* With a finely tuned hot water system you can design a batch water heater with less water storage capacity, thus saving both labor and materials. Or, if your batch is already built, tuning your system will increase the percentage of solar water heating. During warm weather, conservation can make the difference that lets you turn off your auxiliary water heater and go 100% solar.

There are two conservation categories to work on: reduction in the amount of hot water used, and reduction of energy used to heat the water. Savings in the first category certainly helps the second. Because conservation can be achieved using low-cost devices, *conservation is generally more cost-effective than adding solar* and therefore should be done before or along with any solar installation. Most conservation devices pay for themselves in less than 3 years and go on to save money for many more years to come. That is better than 30% interest on your money invested.

HOT WATER

The average American uses between 15 and 30 gal. of hot water per day. Reducing the amount of hot water used is mostly a matter of changing the hardware on your water system. Two major areas of concern are the shower and sink faucets. Changing habits, such as using cold or warm water to wash clothes and taking quick showers, is another area to save hot water. With proper tuning, hot water consumption should be reduced to 10 to 15 gal. a day per person.

The first step is to measure the amount of water that flows from your shower and sinks. Simply collect water from the fixture for 15 seconds, then measure the volume of water and multiply it by four. This gives you gallons per minute (gpm).

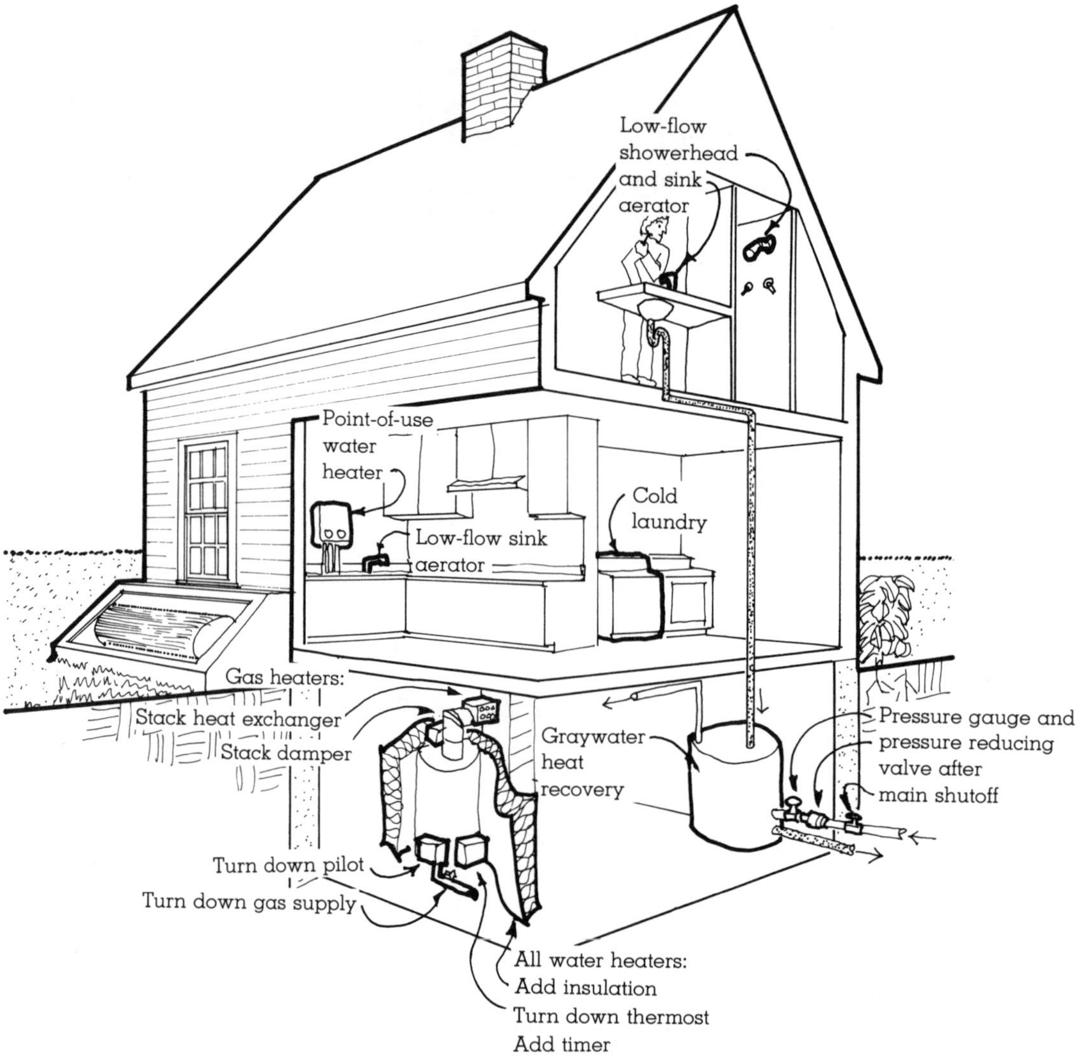

Figure 9–1. Conservation measures cut water-heating costs dramatically.

Conventional showerheads use 4 to 9 gpm, most of which is hot water. With a simple flow-reducing washer costing less than $1 (often free from utility companies), the flow can be reduced to 3 to 5 gpm. But, the shower may be disappointing. Water-saving shower heads, on the other hand, can provide good showers using only 2 to 3 gpm. At 5¢ per kilowatt of electricity, four people showering every other day can save over $100 each year by using a water-saving instead of a conventional shower head. With gas water heating at 50¢ per cubic foot, the savings is about $50 per year.

Because kitchen and bathroom sinks use less water than showers, potential savings are not as great. Typically, sinks without aerators use 5 to 6 gpm. Low-flow aerators pass only 1.5 to 2 gpm and provide adequate rinsing power. You can retrofit existing aerators by adding 50% flow-restricting washers.

Reducing water pressure in your system is another way to reduce water consumption, as well as extend the life of the water tank and reduce water hammering. A pressure-reducing valve and water-pressure gauge can be added to your system for about $50 in parts. Municipal water pressure may be 80 psi or more, and a well pump may be supplying 50 to 75 psi. Many people find that 40 to 50 psi is adequate. By reducing water pressure 50%, flow rates at the fixtures will be reduced by approximately one third. But, reduced water pressure can sometimes leave you with less than satisfying showers. If you do reduce water pressure in the house, leave the outside faucets at full pressure for gardening and in case of fire.

ENERGY

The amount of energy used to supply hot water can be significantly reduced without any change in comfort or life-style. Not that low-flow shower heads and faucets are less than comfortable, but shorter showers and significantly reduced water pressure take some getting used to. These energy-saving techniques range from the no-cost temperature set-back to extensive heat recovery systems.

Most water heaters are factory set to provide 140°F to 160°F water. Shower and washing water is usually below 115°F. By having unnecessarily hot water, heat loss through the tank and pipes is increased. Turning the water heater's thermostat down to 120°F will still provide you with plenty of hot shower water and can save $15 a year (based on 5¢ per kwh). And, with your low-flow shower and faucets you may have even longer hot water usage than before. If your water heater has an upper and lower heating element, set the lower one at least 5°F lower than the upper one.

One potential problem with lower water temperatures is that some dishwashers require slightly hotter water for maximum performance. When purchasing a dishwasher, select one with its own preheater so that using 120°F water is adequate.

The next concern is heat loss through the water tank jacket. Having 120°F water in a 65°F house presents the same temperature difference as a 65°F house when it is 10°F outside. When it is that cold outside (every day) you do not want any less than R-20 insulation. Therefore, you should wrap the tank with 6″ of fiberglass insulation (or the equivalent amount of some other insulation). At 5¢ per kilowatt this added insulation will save about $25 each year.

Note: Electric water heaters can be completely wrapped with insulation. With gas-fired water heaters be careful not to block either the exhaust or intake air passages.

You can further tune your water heater by using a timer control for electric water heaters or a stack damper for gas-fired heaters. By turning your electric water heater on only before periods of greatest hot water demand you can achieve an additional 5% to 10% fuel savings. The stack damper works on gas water heaters by closing the flue that passes through the center of the water heater when the burner is off. When the burner goes on the damper automatically opens. By keeping the flue closed during much of the day, convective heat loss through the flue is reduced.

In some homes, oil-burning furnaces equipped with a domestic water heating coil are used to heat water. When the furnace is being used to heat the house, the water is also heated—a fairly efficient system. But, during warm weather or if you are heating your house without the furnace, with wood or solar for example, then this water-heating system is only 15% to 25% efficient! (Some people heat their houses entirely with wood and are still spending $600 a year for oil.) If you have this type of system, you should design your batch water heater to provide *all* of your hot water during warm weather and turn off your furnace. Any auxiliary water heating that is needed while the furnace is not heating the house should be supplied by a small gas, electric, or tankless water heater.

A tankless water heater is a special type of heater that does not store hot water, but instead heats the water as it is needed. Thus, there is no standby water heat loss, resulting in a significant energy savings. Gas or electric tankless heaters can raise the temperature of incoming water up to 60°F at a flow rate of 1 gal. per minute. This makes them ideal for solar water heater boosters. They can also be used to specifically raise the temperature of dishwasher water. Disadvantages of tankless heaters are that at greater flow rates they may not be able to supply enough hot water and water temperatures may vary slightly while showering. When the tankless heater is used as a solar backup, it is important to use a tankless that is thermostatically controlled rather than one that raises the water temperature by a fixed amount.

Insulation of hot water pipes provides another 3%-5% energy savings. Although this does not add up to much money, the fact that the insulation costs only a few dollars, is easy to install, and lasts indefinitely makes it a worthwhile investment.

Gray Water

Gray water heat recovery is an area for significant energy savings. Gray water is the water that goes down the drain after you use it. Recycling some of the heat in the hot water that goes down the drain can save 20% to 40% of the energy needed to heat the cold water to 120°F. Unfortunately, commercially built heat recovery systems are not yet available in this country, although they are used in Sweden. You can construct your own system (if your plumbing code allows it) from available hardware and plumbing parts.

The potential for gray water heat recovery depends on the volume, temperature, and time coincidence with the hot water demand. The high volume and temperature of hot water from the shower, bath, dishwasher, and clothes washer make them ideal for heat recovery. Because sinks generally use little water, unless they are used to wash dishes, their potential for gray water heat recovery is not great. Gray water characteristics are summarized below.

Gray Water Characteristics

Source	*Volume/Use gallons (liters)*	*Flow Rate gallons (liters)*	*°F*	*(°C)*	*Quality*	*Coincident*
Kitchen sink	up to 5 (1–20)	2.5 (10)/min.	85	(30)	poor	no
Lavatory	up to 1.5 (1–5)	1.25 (5)/min.	85	(30)	fair	no
Bath	32 (120)	5 (20)/min.	100	(37)	good	no
Shower	15 (50)	2 (7)/min.	100	(37)	good	yes
Washer	40 (150)	7 (25)/min.	60	(15)	moot	no

Source: Glenn Nelson, "Greywater Heat Recovery," *Solar Age*, August 1981. Reprinted with permission.

There are two types of gray water heat recovery systems: storage tank and in-line counterflow. The time difference between hot water demand and warm gray water availability determines which system is best.

The storage-tank system is ideal when hot water demand and warm gray water availability are not simultaneous. This system consists of a 25- to 50-gallon insulated holding tank with a heat exchanger containing cold potable water. The insulated tank retains the warm gray water and the heat exchanger transfers the heat to the cold potable water as it passes through. When the hot water demand and warm gray water availability are simultaneous, such as with showering, the in-line counterflow system is more efficient. The in-line counterflow system consists of a heat exchanger inside a large diameter drain pipe.

Note: All gray water heat recovery systems should use double-wall construction with positive leak detection between the gray water and potable water.

Glenn Nelson of Clivus Multrum Northern Rockies in Whitefish, Montana, has constructed both types of systems and measured their performance. A 5′-long in-line counterflow system was connected to a shower that used 2 gal. of water per minute. The heat exchanger was able to raise the temperature of the incoming

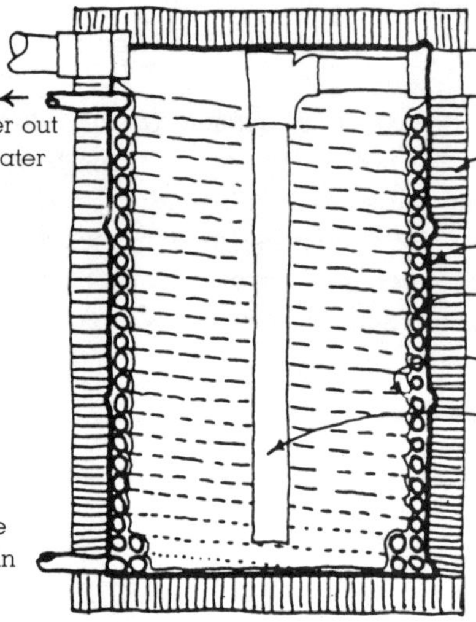

Figure 9–2. Storage tank heat exchanger. To solve the problem of floating scum, the center downspout was modified in a later design to have a removable section at the top, which allows drainage from the surface of the graywater tank. Another idea might be to place a valve (hose bib) opening into the space between the drum wall and the liner. Leaks in the liner could be checked by opening this drain.

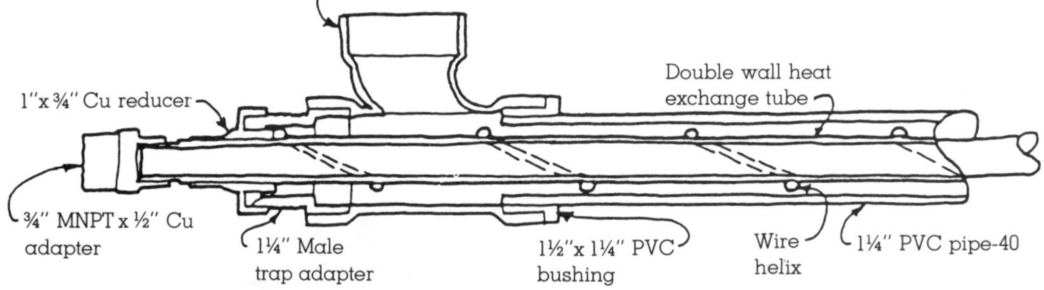

Figure 9–3. In-line Counterflow heat exchanger.

water by 17°F, providing 17% of the hot water energy requirements. It is estimated that a 10′-long exchanger could supply 30% of the hot water energy requirements for showers.

A storage tank system containing 27 gal. of gray water and 100′ of 1″ polybutylene pipe was also tested. It provided 25% of the energy needed for a 9-gal. shower, and only 10% of the energy for a second 9-gal. shower. This emphasizes the value of the in-line counterflow system for shower gray water heat recovery. For other fixture uses, the storage system was able to provide up to 40% energy savings.

Nelson is currently working to produce a single gray water heat recovery system with the advantages of both the storage and in-line systems.

Listed below are the materials costs for each of the systems already tested.

Storage-Tank Heat Exchanger Materials and Costs	
One 55-gal. drum, used	$5.00
One 55-gal. drum liner	5.00
One 55-gal. drum lid	4.00
2″ PVC DWV (4′)	1.50
Miscellaneous PVC fittings	7.00
Miscellaneous hose clamps, adapters	5.50
1″ polybutylene pipe (100′)	50.00
2″ polystyrene insulation (8 sq. ft.)	5.00
2″ polyurethane foam insulation (14 sq. ft.)	10.00
Total	$93.00

In-line Counterflow Heat Exchanger Materials and Costs	
Noranda heat exchange tube #7563LDDA (5′)	$41.50
1 1/4″ PVC pipe, schedule 40 (5′)	1.00
Miscellaneous PVC tees, adapters,	7.00
Miscellaneous copper adapters, couplings	2.00
1 1/2″ fiberglass duct insulation 6 sq. ft.)	1.50
Total	$53.00

Source: Glenn Nelson, "Greywater Heat Recovery," *Solar Age*, August 1981. Reprinted with permission.

APPENDIX

MATERIALS SOURCES

Berry Solar Products, P.O. Box 327, Edison, NJ 08817 (selective surface material, etc.).

Conservation Concepts, Inc., 484 Middletown Road, Hummelstown, PA 17036 (batch heater kits, stainless steel tanks).

Energy House, P.O. Box 5288, Salem, OR 97304; catalog $3.

Energy Shack, P.O. Box 7305, Flint, MI 48507.

Enersorb, Desoto, Inc., 1700 South Mt. Prospect Road, Des Plaines, IL 60018 (selective surfaces).

Integral Design, 4708 Raley Boulevard, Sacramento, CA 95838.

Maxorb, Engenics, 681 Lawlins Road, Wyckoff, NJ 07481 (selective surfaces).

Nature's Way Energy Systems, Route 32, Keene, NH 03431 (batch heater kit, stainless steel tanks, glazing materials, Thermoply, selective surfaces, etc.).

Solar Components Corp., P.O. Box 237, Manchester, NH 03105; catalog $3 (batch heater kit, glass-lined 30-gal. tanks, glazing materials, selective surfaces, etc.).

Solar Hardware Supply Company, 2160 Clay Street, Denver, CO 80211; catalog $1.

Solar Usage Now, Inc., P.O. Box 306, 420 East Tiffin Street, Bascom, OH 44809; catalog $5 (stainless steel tanks, glazing materials, selective surfaces, etc.).

APPENDIX 2

INSULATING VALUES OF MATERIALS

R-Values of Building Materials

Material and Description	*R-Value*
Building Paper:	
Vapor-permeable felt	0.06
Vapor-seal plastic film	negligible
Insulating materials:	
Cellulose	3.1 to 3.8 per inch of thickness
Fiberglass batts	3.1 to 3.4 per inch of thickness
Polystyrene, molded beads (density 1 lb./ft.)	3.6 to 4.0 per inch of thickness
Polystyrene, extruded (Dow Styrofoam, also called blueboard)	5.0 to 5.5 per inch of thickness
Polyurethane, rigid board	5.9 to 7.2 per inch of thickness
Siding materials:	
Shingles	
Asbestos-cement	0.2
Wood, 16″ with 7 1/2″ exposure	0.8
Wood, double 16″ with 12″ exposure	1.2
Siding	
Wood, drop (1″ x 8″)	0.8
Wood, drop (1/2″ x 8″ lapped)	0.8
Wood, bevel (3/4″ x 10″, lapped)	1.1
Plywood, lapped (3/8″)	0.6
Plywood	
1/4″	0.3
3/8″	0.5
1/2″	0.6
5/8″	0.8
3/4″	0.9
Woods	
Hardwoods (maple, oak)	0.9
Softwoods (fir, pine)	1.3

R-Values of Air Films

Type and Orientation of Air Film	Direction of Heat Flow	R-value for Air Film on:		
		Nonreflective surface	*Fairly reflective surface*	*Highly reflective surface*
Still air:				
Horizontal	up	0.61	1.10	1.32
Horizontal	down	0.92	2.70	4.55
45° slope	up	0.62	1.14	1.37
45° slope	down	0.76	1.67	2.22
Vertical	across	0.68	1.35	1.70
Moving air:				
15 mph wind	any*	0.17	-	-
7 1/2 mph wind	any†	0.25	-	-

*Winter conditions.

†Summer conditions.

Source: ASHRAE, *Handbook of Fundamentals,* 1972. Reprinted with permission.

U-Values of Windows

Description	*Winter U-values*[1]
Vertical panels:	
Single pane flat glass	1.13
Insulating glass—double[2]	
3/16″ air space	0.69
1/4″ air space	0.65
1/2″ air space	0.58
Insulating glass—triple[2]	
1/4″ air spaces	0.47
1/2″ air spaces	0.36
Storm windows	
1-4″ air space	0.56
Single plastic sheet	1.09

[1]in units of Btu/hr × ft^2 × °F

[2]double and triple refer to the number of layers of glass

Sources: ASHRAE, *Handbook of Fundamentals,* 1972. Reprinted with permission.

APPENDIX 3

REFLECTANCE, ABSORPTANCE, AND EMITTANCE OF VARIOUS MATERIALS

Material	*% Reflectance*	*% Absorptance*	*Emittance*
fresh snow	87	13	.82
aluminum foil	85	15	.05
white paint	80	20	.91
green paint	50	50	.90
red brick	45	55	.92
concrete	40	60	.97
galvanized iron, new	35	65	.13
grass	31	69	.90
red paint	26	74	.90
water	6	94	.95
flat black paint	4	96	.90
*nickel, black or polished nickel	7	93	.02

*Selective surface

Source: ASHRAE, Handbook of Fundamentals, 1972. Reprinted with permission.

APPENDIX 4

DEGREE DAYS AND DESIGN TEMPERATURES

State	City	Avg. Winter Temp	Design Temp	Sep	Oct	Nov	Dec	Jan	Feb	Mar	Apr	May	Yearly Total
Ala.	Birmingham	54.2	19	6	93	363	555	592	462	363	108	9	2551
	Huntsville	51.3	13	12	127	426	663	694	557	434	138	19	3070
	Mobile	59.9	26	0	22	213	357	415	300	211	42	0	1560
	Montgomery	55.4	22	0	68	330	527	543	417	316	90	0	2291
Alaska	Anchorage	23.0	−25	516	930	1284	1572	1631	1316	1293	879	592	10864
	Fairbanks	6.7	−53	642	1203	1833	2254	2359	1901	1739	1068	555	14279
	Juneau	32.1	− 7	483	725	921	1135	1237	1070	1073	810	601	9075
	Nome	13.1	−32	693	1094	1455	1820	1879	1666	1770	1314	930	14171
Ariz.	Flagstaff	35.6	0	201	558	867	1073	1169	991	911	651	437	7152
	Phoenix	58.5	31	0	22	234	415	474	328	217	75	0	1765
	Tucson	58.1	29	0	25	231	406	471	344	242	75	6	1800
	Winslow	43.0	9	6	245	711	1008	1054	770	601	291	96	4782
	Yuma	64.2	37	0	0	108	264	307	190	90	15	0	974
Ark.	Fort Smith	50.3	9	12	127	450	704	781	596	456	144	22	3292
	Little Rock	50.5	19	9	127	465	716	756	577	434	126	9	3219
	Texarkana	54.2	22	0	78	345	561	626	468	350	105	0	2533
Calif.	Bakersfield	55.4	31	0	37	282	502	546	364	267	105	19	2122
	Burbank	58.6	36	6	43	177	301	366	277	239	138	81	1646
	Eureka	49.9	32	258	329	414	499	546	470	505	438	372	4643
	Fresno	53.3	28	0	84	354	577	605	426	335	162	62	2611
	Long Beach	57.8	36	9	47	171	316	397	311	264	171	93	1803
	Los Angeles	57.4	41	42	78	180	291	372	302	288	219	158	2061
	Oakland	53.5	35	45	127	309	481	527	400	353	255	180	2870
	Sacramento	53.9	30	0	56	321	546	583	414	332	178	72	2502
	San Diego	59.5	42	21	43	135	236	298	235	214	135	90	1458
	San Francisco	55.1	42	102	118	231	388	443	336	319	279	239	3001
	Santa Maria	54.3	32	96	146	270	391	459	370	363	282	233	2967
Colo.	Alamosa	29.7	−17	279	639	1065	1420	1476	1162	1020	696	440	8529
	Colorado Springs	37.3	− 1	132	456	825	1032	1128	938	893	582	319	6423
	Denver	37.6	− 2	117	428	819	1035	1132	938	887	558	288	6283
	Grand Junction	39.3	8	30	313	786	1113	1209	907	729	387	146	5641
	Pueblo	40.4	− 5	54	326	750	986	1085	871	772	429	174	5462

State	City	Avg. Winter Temp	Design Temp	Sep	Oct	Nov	Dec	Jan	Feb	Mar	Apr	May	Yearly Total
Conn.	Bridgeport	39.9	4	66	307	615	986	1079	966	853	510	208	5617
	Hartford	37.3	1	117	394	714	1101	1190	1042	908	519	205	6235
	New Haven	39.0	5	87	347	648	1011	1097	991	871	543	245	5897
Del.	Wilmington	42.5	12	51	270	588	927	980	874	735	387	112	4930
D.C.	Washington	45.7	16	33	217	519	834	871	762	626	288	74	4224
Fla.	Daytona Beach	64.5	32	0	0	75	211	248	190	140	15	0	879
	Fort Myers	68.6	38	0	0	24	109	146	101	62	0	0	442
	Jacksonville	61.9	29	0	12	144	310	332	246	174	21	0	1239
	Key West	73.1	55	0	0	0	28	40	31	9	0	0	108
	Lakeland	66.7	35	0	0	57	164	195	146	99	0	0	661
	Miami	71.1	44	0	0	0	65	74	56	19	0	0	214
	Miami Beach	72.5	45	0	0	0	40	56	36	9	0	0	141
	Orlando	65.7	33	0	0	72	198	220	165	105	6	0	766
	Pensacola	60.4	29	0	19	195	353	400	277	183	36	0	1463
	Tallahassee	60.1	25	0	28	198	360	375	286	202	36	0	1485
	Tampa	66.4	36	0	0	60	171	202	148	102	0	0	683
	West Palm Beach	68.4	40	0	0	6	65	87	64	31	0	0	253
Ga.	Athens	51.8	17	12	115	405	632	642	529	431	141	22	2929
	Atlanta	51.7	18	18	124	417	648	636	518	428	147	25	2961
	Augusta	54.5	20	0	78	333	552	549	445	350	90	0	2397
	Columbus	54.8	23	0	87	333	543	552	434	338	96	0	2383
	Macon	56.2	23	0	71	297	502	505	403	295	63	0	2136
	Rome	49.9	16	24	161	474	701	710	577	468	177	34	3326
	Savannah	57.8	24	0	47	246	437	437	353	254	45	0	1819
Hawaii	Hilo	71.9	59	0	0	0	0	0	0	0	0	0	0
	Honolulu	74.2	60	0	0	0	0	0	0	0	0	0	0
Idaho	Boise	39.7	4	132	415	792	1017	1113	854	722	438	245	5809
	Lewiston	41.0	6	123	403	756	933	1063	815	694	426	239	5542
	Pocatello	34.8	− 8	172	493	900	1166	1324	1058	905	555	319	7033
Ill.	Chicago	37.5	− 4	81	326	753	1113	1209	1044	890	480	211	6155
	Moline	36.4	− 7	99	335	774	1181	1314	1100	918	450	189	6408
	Peoria	38.1	− 2	87	326	759	1113	1218	1025	849	426	183	6025
	Rockford	34.8	− 7	114	400	837	1221	1333	1137	961	516	236	6830
	Springfield	40.6	− 1	72	291	696	1023	1135	935	769	354	136	5429
Ind.	Evansville	45.0	6	66	220	606	896	955	767	620	237	68	4435
	Fort Wayne	37.3	0	105	378	783	1135	1178	1028	890	471	189	6205
	Indianapolis	39.6	0	90	316	723	1051	1113	949	809	432	177	5699
	South Bend	36.6	− 2	111	372	777	1125	1221	1070	933	525	239	6439
Iowa	Burlington	37.6	− 4	93	322	768	1135	1259	1042	859	426	177	6114
	Des Moines	35.5	− 7	96	363	828	1225	1370	1137	915	438	180	6588
	Dubuque	32.7	−11	156	450	906	1287	1420	1204	1026	546	260	7376
	Sioux City	34.0	−10	108	369	867	1240	1435	1198	989	483	214	6951
	Waterloo	32.6	−12	138	428	909	1296	1460	1221	1023	531	229	7320

State	City	Avg. Winter Temp	Design Temp	Sep	Oct	Nov	Dec	Jan	Feb	Mar	Apr	May	Yearly Total
Kans.	Dodge City	42.5	3	33	251	666	939	1051	840	719	354	124	4986
	Goodland	37.8	− 2	81	381	810	1073	1166	955	884	507	236	6141
	Topeka	41.7	3	57	270	672	980	1122	893	722	330	124	5182
	Wichita	44.2	5	33	229	618	905	1023	804	645	270	87	4620
Ky.	Covington	41.4	3	75	291	669	983	1035	893	756	390	149	5265
	Lexington	43.8	6	54	239	609	902	946	818	685	325	105	4683
	Louisville	44.0	8	54	248	609	890	930	818	682	315	105	4660
La.	Alexandria	57.5	25	0	56	273	431	471	361	260	60	0	1921
	Baton Rouge	59.8	25	0	31	216	369	409	294	208	33	0	1560
	Lake Charles	60.5	29	0	19	210	341	381	274	195	39	0	1459
	New Orleans	61.0	32	0	19	192	322	363	258	192	39	0	1385
	Shreveport	56.2	22	0	47	297	477	552	426	304	81	0	2184
Me.	Caribou	24.4	−18	336	682	1044	1535	1690	1470	1308	858	468	9767
	Portland	33.0	− 5	195	508	807	1215	1339	1182	1042	675	372	7511
Md.	Baltimore	43.7	12	48	264	585	905	936	820	679	327	90	4654
	Frederick	42.0	7	66	307	624	955	995	876	741	384	127	5087
Mass.	Boston	40.0	6	60	316	603	983	1088	972	846	513	208	5634
	Pittsfield	32.6	− 5	219	524	831	1231	1339	1196	1063	660	326	7578
	Worcester	34.7	− 3	147	450	774	1172	1271	1123	998	612	304	6969
Mich.	Alpena	29.7	− 5	273	580	912	1268	1404	1299	1218	777	446	8506
	Detroit	37.2	4	87	360	738	1088	1181	1058	936	522	220	6232
	Escanaba	29.6	− 7	243	539	924	1293	1445	1296	1203	777	456	8481
	Flint	33.1	− 1	159	465	843	1212	1330	1198	1066	639	319	7377
	Grand Rapids	34.9	2	135	434	804	1147	1259	1134	1011	579	279	6894
	Lansing	34.8	2	138	431	813	1163	1262	1142	1011	579	273	6909
	Marquette	30.2	− 8	240	527	936	1268	1411	1268	1187	771	468	8393
	Muskegon	36.0	4	120	400	762	1088	1209	1100	995	594	310	6696
	Sault Ste. Marie	27.7	−12	279	580	951	1367	1525	1380	1277	810	477	9048
Minn.	Duluth	23.4	−19	330	632	1131	1581	1745	1518	1355	840	490	10000
	Minneapolis	28.3	−14	189	505	1014	1454	1631	1380	1166	621	288	8382
	Rochester	28.8	−17	186	474	1005	1438	1593	1366	1150	630	301	8295
Miss.	Jackson	55.7	21	0	65	315	502	546	414	310	87	0	2239
	Meridian	55.4	20	0	81	339	518	543	417	310	81	0	2289
	Vicksburg	56.9	23	0	53	279	462	512	384	282	69	0	2041
Mo.	Columbia	42.3	2	54	251	651	967	1076	874	716	324	121	5046
	Kansas City	43.9	4	39	220	612	905	1032	818	682	294	109	4711
	St. Joseph	40.3	− 1	60	285	708	1039	1172	949	769	348	133	5484
	St. Louis	43.1	4	60	251	627	936	1026	848	704	312	121	4900
	Springfield	44.5	5	45	223	600	877	973	781	660	291	105	4900
Mont.	Billings	34.5	−10	186	487	897	1135	1296	1100	970	570	285	7049
	Glasgow	26.4	−25	270	608	1104	1466	1711	1439	1187	648	335	8996
	Great Falls	32.8	−20	258	543	921	1169	1349	1154	1063	642	384	7750
	Havre	28.1	−22	306	595	1065	1367	1584	1364	1181	657	338	8700

State	City	Avg. Winter Temp	Design Temp	Sep	Oct	Nov	Dec	Jan	Feb	Mar	Apr	May	Yearly Total
	Helena	31.1	−17	294	601	1002	1265	1438	1170	1042	651	381	8129
	Kalispell	31.4	− 7	321	654	1020	1240	1401	1134	1029	639	397	8191
	Miles City	31.2	−19	174	502	972	1296	1504	1252	1057	579	276	7723
	Missoula	31.5	− 7	303	651	1035	1287	1420	1120	970	621	391	8125
Neb.	Grand Island	36.0	− 6	108	381	834	1172	1314	1089	908	462	211	6530
	Lincoln	38.8	− 4	75	301	726	1066	1237	1016	834	402	171	5864
	Norfolk	34.0	−11	111	397	873	1234	1414	1179	983	498	233	6979
	North Platte	35.5	− 6	123	440	885	1166	1271	1039	930	519	248	6684
	Omaha	35.6	− 5	105	357	828	1175	1355	1126	939	465	208	6612
	Scottsbluff	35.9	− 8	138	459	876	1128	1231	1008	921	552	285	6673
Nev.	Elko	34.0	−13	225	561	924	1197	1314	1036	911	621	409	7433
	Ely	33.1	− 6	234	592	939	1184	1308	1075	977	672	456	7733
	Las Vegas	53.5	23	0	78	387	617	688	487	335	111	6	2709
	Reno	39.3	2	204	490	801	1026	1073	823	729	510	357	6332
	Winnemucca	36.7	1	210	536	876	1091	1172	916	837	573	363	6761
N.H.	Concord	33.0	−11	177	505	822	1240	1358	1184	1032	636	298	7383
N.J.	Atlantic City	43.2	14	39	251	549	880	936	848	741	420	133	4812
	Newark	42.8	11	30	248	573	921	983	876	729	381	118	4589
	Trenton	42.4	12	57	264	576	924	989	885	753	399	121	4980
N.M.	Albuquerque	45.0	14	12	229	642	868	930	703	595	288	81	4348
	Raton	38.1	− 2	126	431	825	1048	1116	904	834	543	301	6228
	Roswell	47.5	16	18	202	573	806	840	641	481	201	31	3793
	Silver City	48.0	14	6	183	525	729	791	605	518	261	87	3705
N.Y.	Albany	34.6	− 5	138	440	777	1194	1311	1156	992	564	239	6875
	Binghamton	36.6	− 2	141	406	732	1107	1190	1081	949	543	229	6451
	Buffalo	34.5	3	141	440	777	1156	1256	1145	1039	645	329	7062
	New York	42.8	11	30	233	540	902	986	885	760	408	118	4871
	Rochester	35.4	2	126	415	747	1125	1234	1123	1014	597	279	6748
	Schenectady	35.4	− 5	123	422	756	1159	1283	1131	970	543	211	6650
	Syracuse	35.2	− 2	132	415	744	1153	1271	1140	1004	570	248	6756
N.C.	Asheville	46.7	13	48	245	555	775	784	683	592	273	87	4042
	Charlotte	50.4	18	6	124	438	691	691	582	481	156	22	3191
	Greensboro	47.5	14	33	192	513	778	784	672	552	234	47	3805
	Raleigh	49.4	16	21	164	450	716	725	616	487	180	34	3393
	Wilmington	54.6	23	0	74	291	521	546	462	357	96	0	2347
	Winston-Salem	48.4	14	21	171	483	747	753	652	524	207	37	3595
N.D.	Bismarck	26.6	−24	222	577	1083	1463	1708	1442	1203	645	329	8851
	Devils Lake	22.4	−23	273	642	1191	1634	1872	1579	1345	753	381	9901
	Fargo	24.8	−22	219	574	1107	1569	1789	1520	1262	690	332	9226
	Williston	25.2	−21	261	601	1122	1513	1758	1473	1262	681	357	9243
Ohio	Akron-Canton	38.1	1	96	381	726	1070	1138	1016	871	489	202	6037
	Cincinnati	45.1	8	39	208	558	862	915	790	642	294	96	4410
	Cleveland	37.2	2	105	384	738	1088	1159	1047	918	552	260	6351

State	City	Avg. Winter Temp	Design Temp	Sep	Oct	Nov	Dec	Jan	Feb	Mar	Apr	May	Yearly Total
	Columbus	39.7	2	84	347	714	1039	1088	949	809	426	171	5660
	Dayton	39.8	0	78	310	696	1045	1097	955	809	429	167	5622
	Mansfield	36.9	1	114	397	768	1110	1169	1042	924	543	245	6403
	Toledo	36.4	1	117	406	792	1138	1200	1056	924	543	242	6494
	Youngstown	36.8	1	120	412	771	1104	1169	1047	921	540	248	6417
Okla.	Oklahoma City	48.3	11	15	164	498	766	868	664	527	189	34	3725
	Tulsa	47.7	12	18	158	522	787	893	683	539	213	47	3860
Ore.	Astoria	45.6	27	210	375	561	679	753	622	636	480	363	5186
	Eugene	45.6	22	129	366	585	719	803	627	589	426	279	4726
	Medford	43.2	21	78	372	678	871	918	697	642	432	242	5008
	Pendleton	42.6	3	111	350	711	884	1017	773	617	396	205	5127
	Portland	45.6	21	114	335	597	735	825	644	586	396	245	4635
	Roseburg	46.3	25	105	329	567	713	766	608	570	405	267	4491
	Salem	45.4	21	111	338	594	729	822	647	611	417	273	4754
Pa.	Allentown	38.9	3	90	353	693	1045	1116	1002	849	471	167	5810
	Erie	36.8	7	102	391	714	1063	1169	1081	973	585	288	6451
	Harrisburg	41.2	9	63	298	648	992	1045	907	766	396	124	5251
	Philadelphia	41.8	11	60	297	620	965	1016	889	747	392	118	5144
	Pittsburgh	38.4	5	105	375	726	1063	1119	1002	874	480	195	5987
	Reading	42.4	6	54	257	597	939	1001	885	735	372	105	4945
	Scranton	37.2	2	132	434	762	1104	1156	1028	893	498	195	6254
	Williamsport	38.5	1	111	375	717	1073	1122	1002	856	468	177	5934
R.I.	Providence	38.8	6	96	372	660	1023	1110	988	868	534	236	5954
S.C.	Charleston	57.9	26	0	34	210	425	443	367	273	42	0	1794
	Columbia	54.0	20	0	84	345	577	570	470	357	81	0	2484
	Florence	54.5	21	0	78	315	552	552	459	347	84	0	2387
	Greenville-Spartanburg	51.6	18	6	121	399	651	660	546	446	132	19	2980
S.D.	Huron	28.8	−16	165	508	1014	1432	1628	1355	1125	600	288	8223
	Rapid City	33.4	− 9	165	481	897	1172	1333	1145	1051	615	326	7345
	Sioux Falls	30.6	−14	168	462	972	1361	1544	1285	1082	573	270	7839
Tenn.	Bristol	46.2	11	51	236	573	828	828	700	598	261	68	4143
	Chattanooga	50.3	15	18	143	468	698	722	577	453	150	25	3254
	Knoxville	49.2	13	30	171	489	725	732	613	493	198	43	3494
	Memphis	50.5	17	18	130	447	698	729	585	456	147	22	3232
	Nashville	48.9	12	30	158	495	732	778	644	512	189	40	3578
Tex.	Abilene	53.9	17	0	99	366	586	642	470	347	114	0	2624
	Amarillo	47.0	8	18	205	570	797	877	664	546	252	56	3985
	Austin	59.1	25	0	31	225	388	468	325	223	51	0	1711
	Corpus Christi	64.6	32	0	0	120	220	291	174	109	0	0	914
	Dallas	55.3	19	0	62	321	524	601	440	319	90	6	2363
	El Paso	52.9	21	0	84	414	648	685	445	319	105	0	2700
	Galveston	62.2	32	0	6	147	276	360	263	189	33	0	1274

State	City	Avg. Winter Temp	Design Temp	Sep	Oct	Nov	Dec	Jan	Feb	Mar	Apr	May	Yearly Total
	Houston	61.0	28	0	6	183	307	384	288	192	36	0	1396
	Laredo	66.0	32	0	0	105	217	267	134	74	0	0	797
	Lubbock	48.8	11	18	174	513	744	800	613	484	201	31	3578
	Port Arthur	60.5	29	0	22	207	329	384	274	192	39	0	1447
	San Antonio	60.1	25	0	31	204	363	428	286	195	39	0	1546
	Waco	57.2	21	0	43	270	456	536	389	270	66	0	2030
	Wichita Falls	53.0	15	0	99	381	632	698	518	378	120	6	2832
Utah	Milford	36.5	− 1	99	443	867	1141	1252	988	822	519	279	6497
	Salt Lake City	38.4	5	81	419	849	1082	1172	910	763	459	233	6052
Vt.	Burlington	29.4	−12	207	539	891	1349	1513	1333	1187	714	353	8269
Va.	Lynchburg	46.0	15	51	223	540	822	849	731	605	267	78	4166
	Norfolk	49.2	20	0	136	408	698	738	655	533	216	37	3421
	Richmond	47.3	14	36	214	495	784	815	703	546	219	53	3865
	Roanoke	46.1	15	51	229	549	825	834	722	614	261	65	4150
Wash.	Olympia	44.2	21	198	422	636	753	834	675	645	450	307	5236
	Seattle	46.9	28	129	329	543	657	738	599	577	396	242	4424
	Spokane	36.5	− 2	168	493	879	1082	1231	980	834	531	288	6655
	Walla Walla	43.8	12	87	310	681	843	986	745	589	342	177	4805
	Yakima	39.1	6	144	450	828	1039	1163	868	713	435	220	5941
W. Va.	Charleston	44.8	9	63	254	591	865	880	770	648	300	96	4476
	Elkins	40.1	1	135	400	729	992	1008	896	791	444	198	5675
	Huntington	45.0	10	63	257	585	856	880	764	636	294	99	4446
	Parkersburg	43.5	8	60	264	606	905	942	826	691	339	115	4754
Wisc.	Green Bay	30.3	−12	174	484	924	1333	1494	1313	1141	654	335	8029
	La Crosse	31.5	−12	153	437	924	1339	1504	1277	1070	540	245	7589
	Madison	30.9	− 9	174	474	930	1330	1473	1274	1113	618	310	7863
	Milwaukee	32.6	− 6	174	471	876	1252	1376	1193	1054	642	372	7635
Wyo.	Casper	33.4	−11	192	524	942	1169	1290	1084	1020	657	381	7410
	Cheyenne	34.2	− 6	219	543	909	1085	1212	1042	1026	702	428	7381
	Lander	31.4	−16	204	555	1020	1299	1417	1145	1017	654	381	7870
	Sheridan	32.5	−12	219	539	948	1200	1355	1154	1051	642	366	7680

Source: ASHRAE, *Handbook of Fundamentals, 1972.* Reprinted with permission.

APPENDIX 5

MAP OF THE AVERAGE PERCENTAGE OF THE TIME THE SUN IS SHINING

JANUARY

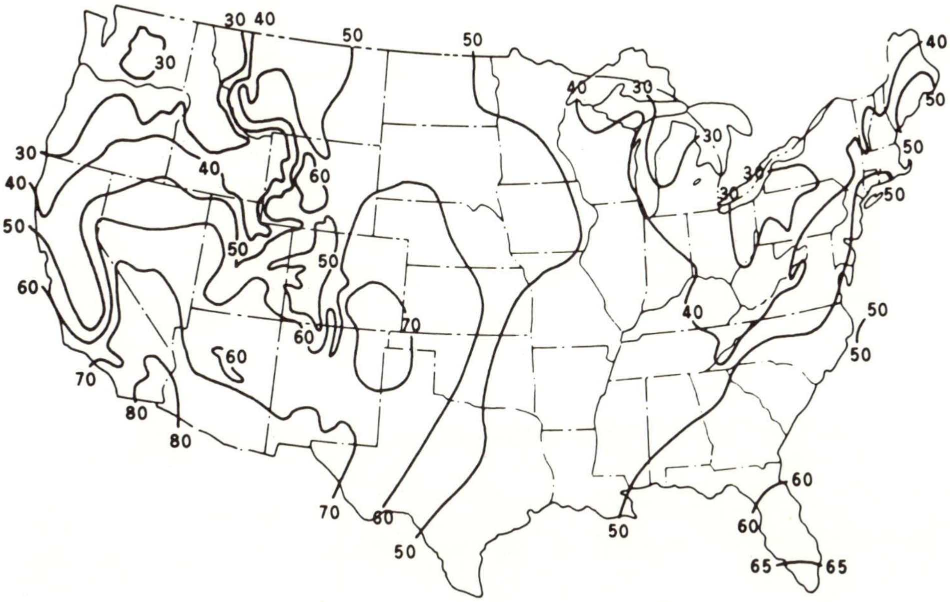

FEBRUARY

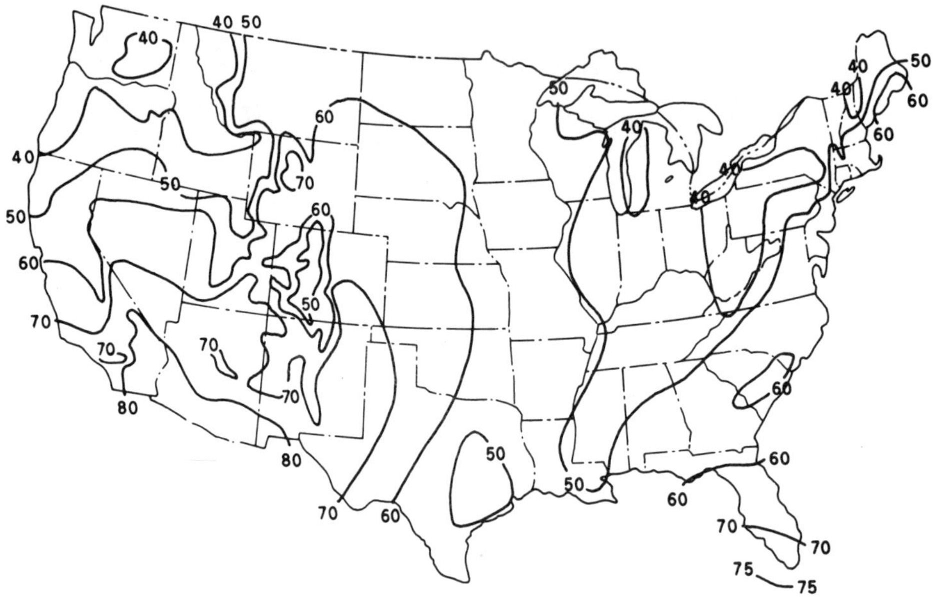

MARCH

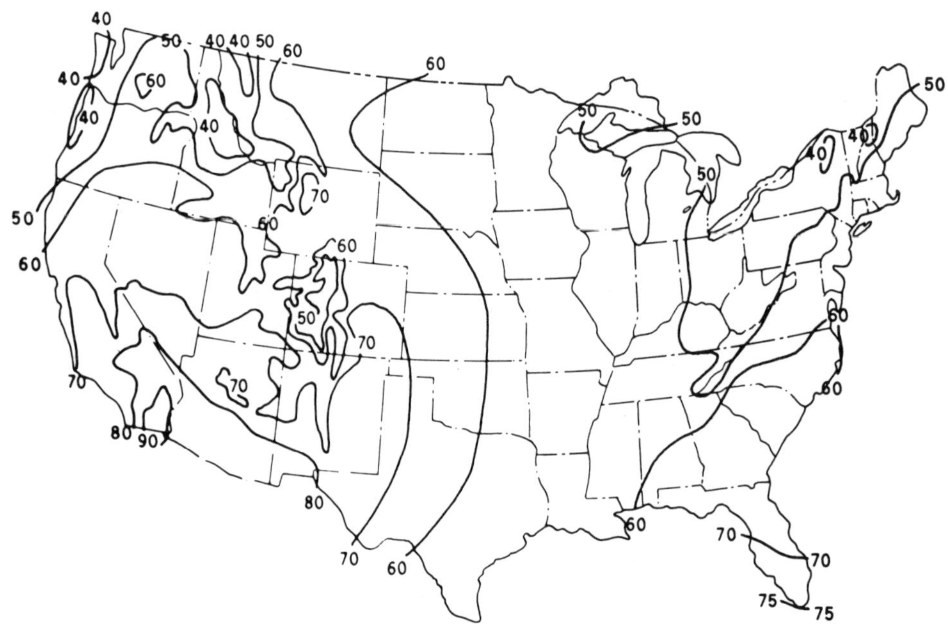

APRIL

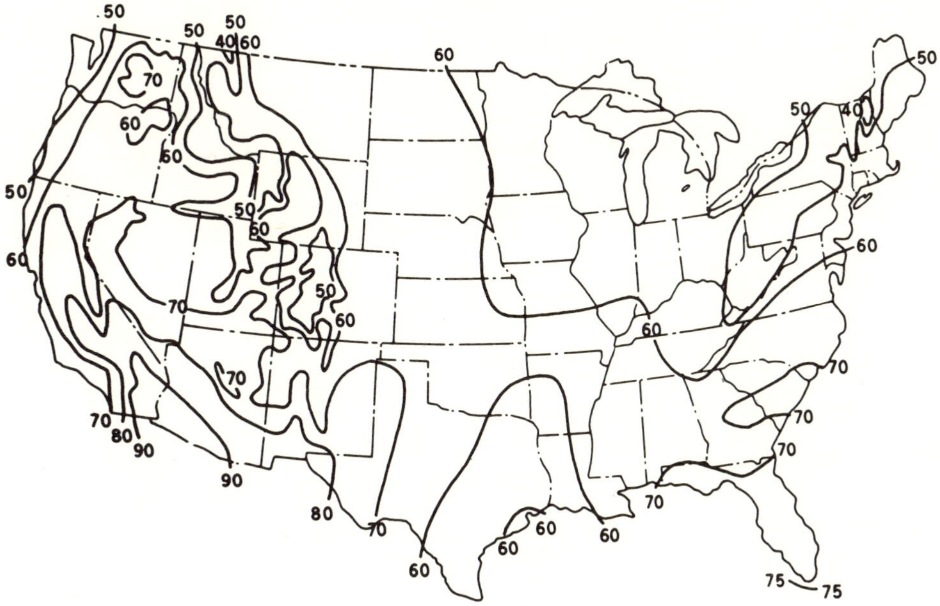
50
50
50
40
60
60
50
70
60
50
50
50
50
60
70
50
60
60
70
70
80
90
90
80
70
60
60
60
60
70
70
70
70
60
40
50
75
75

MAY

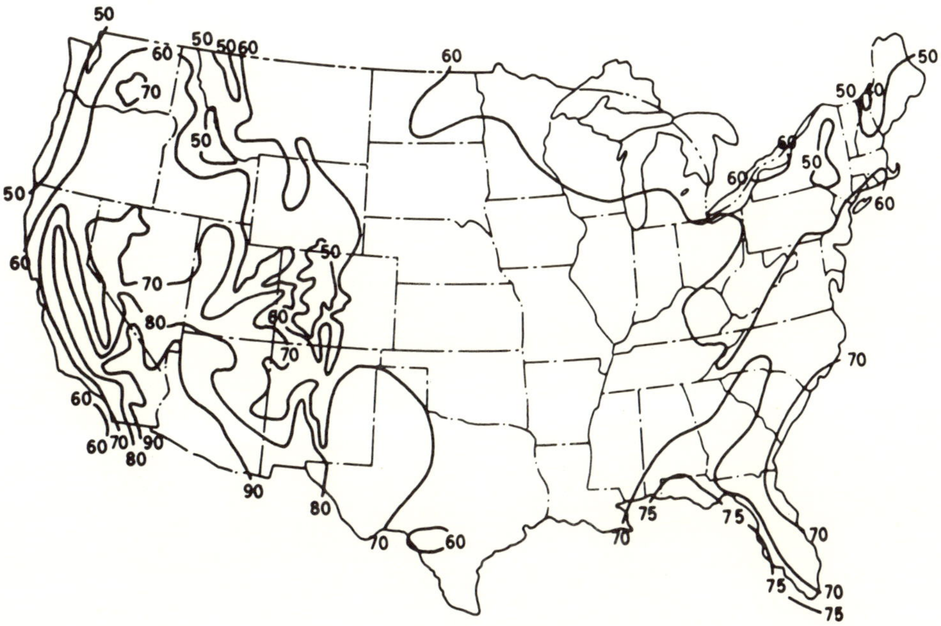
50
60
50
50
60
60
50
70
50
50
50
60
70
50
80
60
70
60
60
70
90
80
90
80
70
60
70
75
75
70
75
70
75
50
40
50
60
60
50
60
70

SEPTEMBER

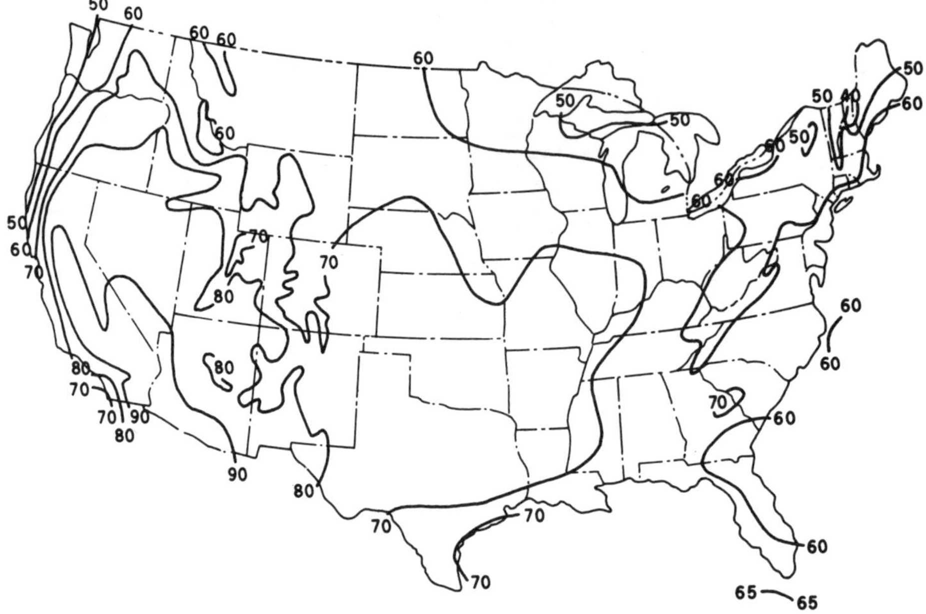

OCTOBER

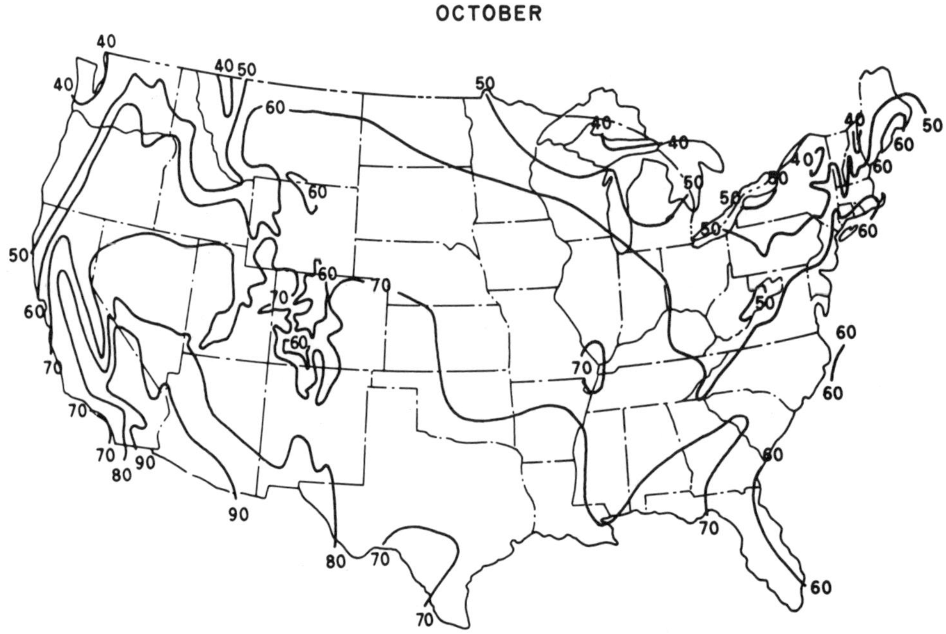

NOVEMBER

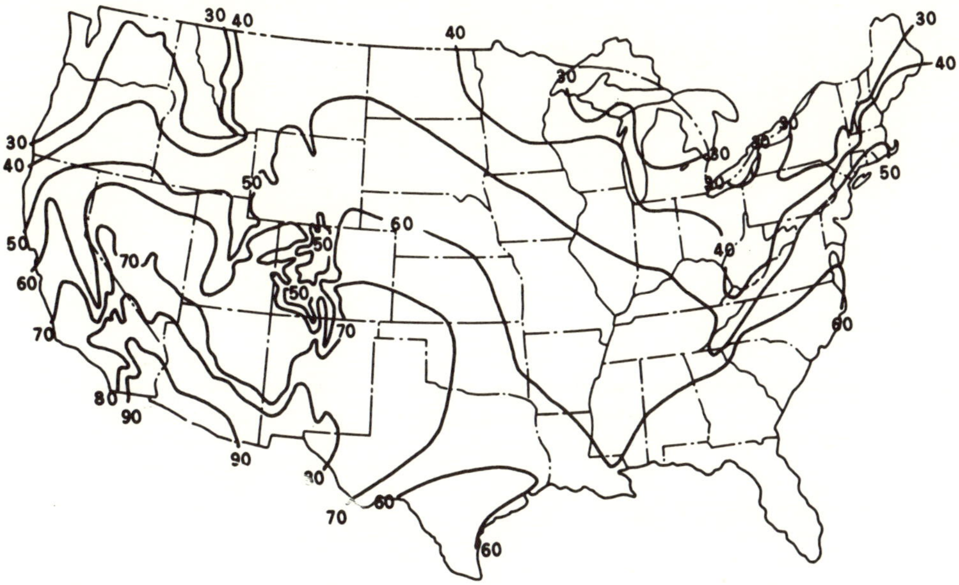

DECEMBER

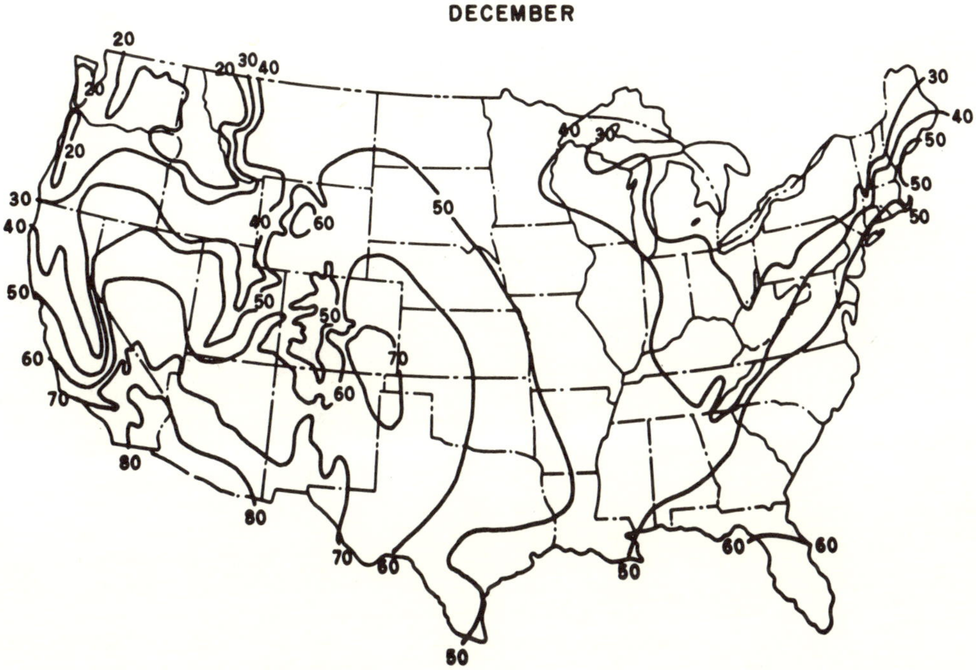

Source: Passive Solar Energy, B. Anderson and M. Wells, Brick House Publishing Co., Inc., 1981.

APPENDIX 6

CLEAR-DAY INSOLATION DATA

Solar Position and Insolation, 24° N Latitude

Date	*Solar Time*		*Solar Position*		*BtuH/Sq. Ft. Total Insolation on Surfaces*					
						South-Facing Surface Angle with Horizon				
	A.M.	*P.M.*	*Alt.*	*Azm.*	*Horiz.*	*14*	*24*	*34*	*54*	*90*
Jan. 21	7	5	4.8	65.6	10	17	21	25	28	31
	8	4	16.9	58.3	83	110	126	137	145	127
	9	3	27.9	48.8	151	188	207	221	228	176
	10	2	37.2	36.1	204	246	268	282	287	207
	11	1	43.6	19.6	237	283	306	319	324	226
	12		46.0	0.0	249	296	319	332	336	232
			Surface Daily Totals		1622	1984	2174	2300	2360	1766
Feb. 21	7	5	9.3	74.6	35	44	49	53	56	46
	8	4	22.3	67.2	116	135	145	150	151	102
	9	3	34.4	57.6	187	213	225	230	228	141
	10	2	45.1	44.2	241	273	286	291	287	168
	11	1	53.0	25.0	276	310	324	328	323	185
	12		56.0	0.0	288	323	337	341	335	191
			Surface Daily Totals		1998	2276	2396	2446	2424	1476
Mar. 21	7	5	13.7	83.8	60	63	64	62	59	27
	8	4	27.2	76.8	141	150	152	149	142	64
	9	3	40.2	67.9	212	226	229	225	214	95
	10	2	52.3	54.8	266	285	288	283	270	120
	11	1	61.9	33.4	300	322	326	320	305	135
	12		66.0	0.0	312	334	339	333	317	140
			Surface Daily Totals		2270	2428	2456	2412	2298	1022

Apr. 21	6	6	4.7	100.6	7	5	4	4	3	2
	7	5	18.3	94.9	83	77	70	62	51	10
	8	4	32.0	89.0	160	157	149	137	122	16
	9	3	45.6	81.9	227	227	220	206	186	41
	10	2	59.0	71.8	278	282	275	259	237	61
	11	1	71.1	51.6	310	316	309	293	269	74
	12		77.6	0.0	321	328	321	305	280	79
	Surface Daily Totals				2454	2458	2374	2228	2016	488
May 21	6	6	8.0	108.4	22	15	10	9	9	5
	7	5	21.2	103.2	98	85	73	59	44	12
	8	4	34.6	98.5	171	159	145	127	106	15
	9	3	48.3	93.6	233	224	210	190	165	16
	10	2	62.0	87.7	281	275	261	239	211	22
	11	1	75.5	76.9	311	307	293	270	240	34
	12		86.0	0.0	322	317	304	281	250	37
	Surface Daily Totals				2556	2447	2286	2072	1800	246
Jun 21	6	6	9.3	111.6	29	20	12	12	11	7
	7	5	22.3	106.8	103	87	73	58	41	13
	8	4	35.5	102.6	173	158	142	122	99	16
	9	3	49.0	98.7	234	221	204	182	155	18
	10	2	62.6	95.0	280	269	253	229	199	18
	11	1	76.3	90.8	309	300	283	259	227	19
	12		89.4	0.0	319	310	294	269	236	22
	Surface Daily Totals				2574	2422	2230	1992	1700	204
Jul 21	6	6	8.2	109.0	23	16	11	10	9	6
	7	5	21.4	103.8	98	85	73	59	44	13
	8	4	34.8	99.2	169	157	143	125	104	16
	9	3	48.4	94.5	231	221	207	187	161	18
	10	2	62.1	89.0	278	270	256	235	206	21
	11	1	75.7	79.2	307	302	287	265	235	32
	12		86.6	0.0	317	312	298	275	245	36
	Surface Daily Totals				2526	2412	2250	2036	1766	246
Aug 21	6	6	5.0	101.3	7	5	4	4	4	2
	7	5	18.5	95.6	82	76	69	60	50	11
	8	4	32.2	89.7	158	154	146	134	118	16
	9	3	45.9	82.9	223	222	214	200	181	39
	10	2	59.3	73.0	273	275	268	252	230	58
	11	1	71.6	53.2	304	309	301	285	261	71
	12		78.3	0.0	315	320	313	296	272	75
	Surface Daily Totals				2408	2402	2316	2168	1958	470

Sep 21	7	5	13.7	83.8	57	60	60	59	56	26
	8	4	27.2	76.8	136	144	146	143	136	62
	9	3	40.2	67.9	205	218	221	217	206	93
	10	2	52.3	54.8	258	275	278	273	261	116
	11	1	61.9	33.4	291	311	315	309	295	131
	12		66.0	0.0	302	323	327	321	306	136
		Surface Daily Totals			2194	2342	2366	2322	2212	992
Oct.21	7	5	9.1	74.1	32	40	45	48	50	42
	8	4	22.0	66.7	111	129	139	144	145	99
	9	3	34.1	57.1	180	206	217	223	221	138
	10	2	44.7	43.8	234	265	277	282	279	165
	11	1	52.5	24.7	268	301	315	319	314	182
	12		55.5	0.0	279	314	328	332	327	188
		Surface Daily Totals			1928	2198	2314	2364	2346	1442
Nov. 21	7	5	4.9	65.8	10	16	20	24	27	29
	8	4	17.0	58.4	82	108	123	135	142	124
	9	3	28.0	48.9	150	186	205	217	224	172
	10	2	37.3	36.3	203	244	265	278	283	204
	11	1	43.8	19.7	236	280	302	316	320	222
	12		46.2	0.0	247	293	315	328	332	228
		Surface Daily Totals			1610	1962	2146	2268	2324	1730
Dec. 21	7	5	3.2	62.6	3	7	9	11	12	14
	8	4	14.9	55.3	71	99	116	129	139	130
	9	3	25.5	46.0	137	176	198	214	223	184
	10	2	34.3	33.7	189	234	258	275	283	217
	11	1	40.4	18.2	221	270	295	312	320	236
	12		42.6	0.0	232	282	308	325	332	243
		Surface Daily Totals			1474	1852	2058	2204	2286	1808

Solar Position and Insolation, 32° N Latitude

					BtuH/Sq. Ft. Total Insolation on Surfaces					
	Solar Time		*Solar Position*			*South-Facing Surface Angle with Horizon*				
Date	*A.M.*	*P.M.*	*Alt.*	*Azm.*	*Horiz.*	*22*	*32*	*42*	*52*	*90*
Jan. 21	7	5	1.4	65.2	0	0	0	0	1	1
	8	4	12.5	56.5	56	93	106	116	123	115
	9	3	22.5	46.0	118	175	193	206	212	181

	10	2	30.6	33.1	167	235	256	269	274	221
	11	1	36.1	17.5	198	273	295	308	312	245
	12		38.0	0.0	209	285	308	321	324	253
		Surface Daily Totals			1288	1839	2008	2118	2166	1779
Feb. 21	7	5	7.1	73.5	22	34	37	40	42	38
	8	4	19.0	64.4	95	127	136	140	141	108
	9	3	29.9	53.4	161	206	217	222	220	158
	10	2	39.1	39.4	212	266	278	283	279	193
	11	1	45.6	21.4	244	304	317	321	315	214
	12		48.0	0.0	255	316	330	334	328	222
		Surface Daily Totals			1724	2188	2300	2345	2322	1644
Mar. 21	7	5	12.7	81.9	54	60	60	59	56	32
	8	4	25.1	73.0	129	146	147	144	137	78
	9	3	36.8	62.1	194	222	224	220	209	119
	10	2	47.3	47.5	245	280	283	278	265	150
	11	1	55.0	26.8	277	317	321	315	300	170
	12		58.0	0.0	287	329	333	327	312	177
		Surface Daily Totals			2084	2378	2403	2358	2246	1276
Apr. 21	6	6	6.1	99.9	14	9	6	6	5	3
	7	5	18.8	92.2	86	78	71	62	51	10
	8	4	31.5	84.0	158	156	148	136	120	35
	9	3	43.9	74.2	220	225	217	203	183	68
	10	2	55.7	60.3	267	279	272	256	234	95
	11	1	65.4	37.5	297	313	306	290	265	112
	12		69.6	0.0	307	325	318	301	276	118
		Surface Daily Totals			2390	2444	2356	2206	1994	764
May 21	6	6	10.4	107.2	36	21	13	13	12	7
	7	5	22.8	100.1	107	88	75	60	44	13
	8	4	35.4	92.9	175	159	145	127	105	15
	9	3	48.1	84.7	233	223	209	188	163	33
	10	2	60.6	73.3	277	273	259	237	208	56
	11	1	72.0	51.9	305	305	290	268	237	72
	12		78.0	0.0	315	315	301	278	247	77
		Surface Daily Totals			2582	2454	2284	2064	1788	469
Jun 21	6	6	12.2	110.2	45	26	16	15	14	9
	7	5	24.3	103.4	115	91	76	59	41	14
	8	4	36.9	96.8	180	159	143	122	99	16
	9	3	49.6	89.4	236	221	204	181	153	19
	10	2	62.2	79.7	279	268	251	227	197	41
	11	1	74.2	60.9	306	299	282	257	224	56

		12	81.5	0.0	315	309	292	267	234	60
		Surface Daily Totals			2634	2436	2234	1990	1690	370
Jul 21	6	6	10.7	107.7	37	22	14	13	12	8
	7	5	23.1	100.6	107	87	75	60	44	14
	8	4	35.7	93.6	174	158	143	125	104	16
	9	3	48.4	85.5	231	220	205	185	159	31
	10	2	60.9	74.3	274	269	254	232	204	54
	11	1	72.4	53.3	302	300	285	262	232	69
		12	78.6	0.0	311	310	296	273	242	74
		Surface Daily Totals			2558	2422	2250	2030	1754	458
Aug. 21	6	6	6.5	100.5	14	9	7	6	6	4
	7	5	19.1	92.8	85	77	69	60	50	12
	8	4	31.8	84.7	156	152	144	132	116	33
	9	3	44.3	75.0	216	220	212	197	178	65
	10	2	56.1	61.3	262	272	264	249	226	91
	11	1	66.0	38.4	292	305	298	281	257	107
		12	70.3	0.0	302	317	309	292	268	113
		Surface Daily Totals			2352	2388	2296	2144	1934	736
Sep. 21	7	5	12.7	81.9	51	56	56	55	52	30
	8	4	25.1	73.0	124	140	141	138	131	75
	9	3	36.8	62.1	188	213	215	211	201	114
	10	2	47.3	47.5	237	270	273	268	255	145
	11	1	55.0	26.8	268	306	309	303	289	164
		12	58.0	0.0	278	318	321	315	300	171
		Surface Daily Totals			2014	2288	2308	2264	2154	1226
Oct. 21	7	5	6.8	73.1	19	29	32	34	36	32
	8	4	18.7	64.0	90	120	128	133	134	104
	9	3	29.5	53.0	155	198	208	213	212	153
	10	2	38.7	39.1	204	257	269	273	270	188
	11	1	45.1	21.1	236	294	307	311	306	209
		12	47.5	0.0	247	306	320	324	318	217
		Surface Daily Totals			1654	2100	2208	2252	2232	1588
Nov. 21	7	5	1.5	65.4	0	0	0	1	1	1
	8	4	12.7	56.6	55	91	104	113	119	111
	9	3	22.6	46.1	118	173	190	202	208	176
	10	2	30.8	33.2	166	233	252	265	270	217
	11	1	36.2	17.6	197	270	291	303	307	241
		12	38.2	0.0	207	282	304	316	320	249
		Surface Daily Totals			1280	1816	1980	2084	2130	1742

Dec. 21	8	4	10.3	53.8	41	77	90	101	108	107
	9	3	19.8	43.6	102	161	180	195	204	183
	10	2	27.6	31.2	150	221	244	259	267	226
	11	1	32.7	16.4	180	258	282	298	305	251
	12		34.6	0.0	190	271	295	311	318	259
		Surface Daily Totals			1136	1704	1888	2016	2086	1794

Solar Position and Insolation, 40° N Latitude

	Solar Time		*Solar Position*		*BtuH/Sq. Ft. Total Insolation on Surfaces*					
						South-Facing Surface Angle with Horizon				
Date	*A.M.*	*P.M.*	*Alt.*	*Azm.*	*Horiz.*	*30*	*40*	*50*	*60*	*90*
Jan. 21	8	4	8.1	55.3	28	65	74	81	85	84
	9	3	16.8	44.0	83	155	171	182	187	171
	10	2	23.8	30.9	127	218	237	249	254	223
	11	1	28.4	16.0	154	257	277	290	293	253
	12		30.0	0.0	164	270	291	303	306	263
		Surface Daily Totals			948	1660	1810	1906	1944	1726
Feb. 21	7	5	4.8	72.7	10	19	21	23	24	22
	8	4	15.4	62.2	73	114	122	126	127	107
	9	3	25.0	50.2	132	195	205	209	208	167
	10	2	32.8	35.9	178	256	267	271	267	210
	11	1	38.1	18.9	206	293	306	310	304	236
	12		40.0	0.0	216	306	319	323	317	245
		Surface Daily Totals			1414	2060	2162	2202	2176	1730
Mar. 21	7	5	11.4	80.2	46	55	55	54	51	35
	8	4	22.5	69.6	114	140	141	138	131	89
	9	3	32.8	57.3	173	215	217	213	202	138
	10	2	41.6	41.9	218	273	276	271	258	176
	11	1	47.7	22.6	247	310	313	307	293	200
	12		50.0	0.0	257	322	326	320	305	208
		Surface Daily Totals			1852	2308	2330	2284	2174	1484
Apr. 21	6	6	7.4	98.9	20	11	8	7	7	4
	7	5	18.9	89.5	87	77	70	61	50	12
	8	4	30.3	79.3	152	153	145	133	117	53

	9	3	41.3	67.2	207	221	213	199	179	93
	10	2	51.2	51.4	250	275	267	252	229	126
	11	1	58.7	29.2	277	308	301	285	260	147
	12		61.6	0.0	287	320	313	296	271	154
			Surface Daily Totals		2274	2412	2320	2168	1956	1022
May 21	5	7	1.9	114.7	0	0	0	0	0	0
	6	6	12.7	105.6	49	25	15	14	13	9
	7	5	24.0	96.6	214	89	76	60	44	13
	8	4	35.4	87.2	175	158	144	125	104	25
	9	3	46.8	76.0	227	221	206	186	160	60
	10	2	57.5	60.9	267	270	255	233	205	89
	11	1	66.2	37.1	293	301	287	264	234	108
	12		70.0	0.0	301	312	297	274	243	114
			Surface Daily Totals		2552	2442	2264	2040	1760	724
Jun 21	5	7	4.2	117.3	4	3	3	2	2	1
	6	6	14.8	108.4	60	30	18	17	16	10
	7	5	26.0	99.7	123	92	77	59	41	14
	8	4	37.4	90.7	182	159	142	121	97	16
	9	3	48.8	80.2	233	219	202	179	151	47
	10	2	59.8	65.8	272	266	248	224	194	74
	11	1	69.2	41.9	296	296	278	253	221	82
	12		73.5	0.0	304	306	289	263	230	98
			Surface Daily Totals		2648	2434	2224	1974	1670	610
Jul 21	5	7	2.3	115.2	0	0	0	0	0	0
	6	6	13.1	106.1	50	26	17	15	14	9
	7	5	24.3	97.2	114	89	75	60	44	14
	8	4	35.8	87.8	174	157	142	124	102	24
	9	3	47.2	76.7	225	218	203	182	157	58
	10	2	57.9	61.7	265	266	251	229	200	86
	11	1	66.7	37.9	290	296	281	258	228	104
	12		70.6	0.0	298	307	292	269	238	111
			Surface Daily Totals		2534	2409	2230	2006	1728	702
Aug. 21	6	6	7.9	99.5	21	12	9	8	7	5
	7	5	19.3	90.0	87	76	69	60	49	12
	8	4	30.7	79.9	150	150	141	129	113	50
	9	3	41.8	67.9	205	216	207	193	173	89
	10	2	51.7	52.1	246	267	259	244	221	120
	11	1	59.3	29.7	273	300	292	276	252	140
	12		62.3	0.0	282	311	303	287	262	147
			Surface Daily Totals		2244	2354	2258	2104	1894	978

Sept. 21	7	5	11.4	80.2	43	51	51	49	47	32
	8	4	22.5	69.6	109	133	134	131	124	84
	9	3	32.8	57.3	167	206	208	203	193	132
	10	2	41.6	41.9	211	262	265	260	247	168
	11	1	47.7	22.6	239	298	301	295	281	192
	12		50.0	0.0	249	310	313	307	292	200
	Surface Daily Totals				1788	2210	2228	2182	2074	1416
Oct. 21	7	5	4.5	72.3	7	14	15	17	17	16
	8	4	15.0	61.9	68	106	113	117	118	100
	9	3	24.5	49.8	126	185	195	200	198	160
	10	2	32.4	35.6	170	245	257	261	257	203
	11	1	37.6	18.7	199	283	295	299	294	229
	12		39.5	0.0	208	295	308	312	306	238
	Surface Daily Totals				1348	1962	2060	2098	2074	1654
Nov. 21	8	4	8.2	55.4	28	63	72	78	82	81
	9	3	17.0	44.1	82	152	167	178	183	167
	10	2	24.0	31.0	126	215	233	245	249	219
	11	1	28.6	16.1	153	254	273	285	288	248
	12		30.2	0.0	163	267	287	298	301	258
	Surface Daily Totals				942	1636	1778	1870	1908	1686
Dec. 21	8	4	5.5	53.0	14	39	45	50	54	56
	9	3	14.0	41.9	65	135	152	164	171	163
	10	2	20.7	29.4	107	200	221	235	242	221
	11	1	25.0	15.2	134	239	262	276	283	252
	12		26.6	0.0	143	253	275	290	296	263
	Surface Daily Totals				782	1480	1634	1740	1796	1646

Solar Position and Insolation, 48° N Latitude

	Solar Time		*Solar Position*		*BtuH/Sq. Ft. Total Insolation on Surfaces*					
						South-Facing Surface Angle with Horizon				
Date	*A.M.*	*P.M.*	*Alt.*	*Azm.*	*Horiz.*	*38*	*48*	*58*	*68*	*90*
Jan. 21	8	4	3.5	54.6	4	17	19	21	22	22
	9	3	11.0	42.6	46	120	132	140	145	139
	10	2	16.9	29.4	83	190	206	216	220	206

	11	1	20.7	15.1	107	231	249	260	263	243
	12		22.0	0.0	115	245	264	275	278	255
			Surface Daily Totals		596	1360	1478	1550	1578	1478
Feb. 21	7	5	2.4	72.2	1	3	4	4	4	4
	8	4	11.6	60.5	49	95	102	105	106	96
	9	3	19.7	47.7	100	178	187	191	190	167
	10	2	26.2	33.3	139	240	251	255	251	217
	11	1	30.5	17.2	165	278	290	294	288	247
	12		32.0	0.0	173	291	304	307	301	258
			Surface Daily Totals		1080	1880	1972	2024	1978	1720
Mar. 21	7	5	10.0	78.7	37	49	49	47	45	35
	8	4	19.5	66.8	96	131	132	129	122	96
	9	3	28.2	53.4	147	205	207	203	193	152
	10	2	35.4	37.8	187	263	266	261	248	195
	11	1	40.3	19.8	212	300	303	297	283	223
	12		42.0	0.0	220	312	315	309	294	232
			Surface Daily Totals		1578	2208	2228	2182	2074	1632
Apr. 21	6	6	8.6	97.8	27	13	9	8	7	5
	7	5	18.6	86.7	85	76	69	59	48	21
	8	4	28.5	74.9	142	149	141	129	113	69
	9	3	37.8	61.2	191	216	208	194	174	115
	10	2	45.8	44.6	228	268	260	245	223	152
	11	1	51.5	24.0	252	301	294	278	254	177
	12		53.6	0.0	260	313	305	289	264	185
			Surface Daily Totals		2106	2358	2266	2114	1902	1262
May 21	5	7	5.2	114.3	9	4	4	4	3	2
	6	6	14.7	103.7	61	27	16	15	13	10
	7	5	24.6	93.0	118	89	75	60	43	13
	8	4	34.7	81.6	171	156	142	123	101	45
	9	3	44.3	68.3	217	217	202	182	156	86
	10	2	53.0	51.3	252	265	251	229	200	120
	11	1	59.5	28.6	274	296	281	258	228	141
	12		62.0	0.0	281	306	292	269	238	149
			Surface Daily Totals		2482	2418	2234	2010	1728	982
Jun 21	5	7	7.9	116.5	21	9	9	8	7	5
	6	6	17.2	106.2	74	33	19	18	16	12
	7	5	27.0	95.8	129	93	77	59	39	15
	8	4	37.1	84.6	181	157	140	119	95	35

	9	3	46.9	71.6	225	216	198	175	147	74
	10	2	55.8	54.8	259	262	244	220	189	105
	11	1	62.7	31.2	280	291	273	248	216	126
	12		65.5	0.0	287	301	283	258	225	133
			Surface Daily Totals		2628	2420	2204	1950	1644	874
Jul 21	5	7	5.7	114.7	10	5	5	4	4	3
	6	6	15.2	104.1	62	28	18	16	15	11
	7	5	25.1	93.5	118	89	75	59	42	14
	8	4	35.1	82.1	171	154	140	121	99	43
	9	3	44.8	68.8	215	214	199	178	153	83
	10	2	53.5	51.9	250	261	246	224	195	116
	11	1	60.1	29.0	272	291	276	253	223	137
	12		62.6	0.0	279	301	286	263	232	144
			Surface Daily Totals		2474	2386	2200	1974	1694	956
Aug 21	6	6	9.1	98.3	28	14	10	9	8	6
	7	5	19.1	87.2	85	75	67	58	47	20
	8	4	29.0	75.4	141	145	137	125	109	65
	9	3	38.4	61.8	189	210	201	187	168	110
	10	2	46.4	45.1	225	260	252	237	214	146
	11	1	52.2	24.3	248	293	285	268	244	169
	12		54.3	0.0	256	304	296	279	255	177
			Surface Daily Totals		2086	2300	2200	2046	1836	1208
Sep. 21	7	5	10.0	78.7	35	44	44	43	40	31
	8	4	19.5	66.8	92	124	124	121	115	90
	9	3	28.2	53.4	142	196	197	193	183	143
	10	2	35.4	37.8	181	251	254	248	236	185
	11	1	40.3	19.8	205	287	289	284	269	212
	12		42.0	0.0	213	299	302	296	281	221
			Surface Daily Totals		1522	2102	2118	2070	1966	1546
Oct. 21	7	5	2.0	71.9	0	1	1	1	1	1
	8	4	11.2	60.2	44	86	91	95	95	87
	9	3	19.3	47.4	94	167	176	180	178	157
	10	2	25.7	33.1	133	228	239	242	239	207
	11	1	30.0	17.1	157	266	277	281	276	237
	12		31.5	0.0	166	279	291	294	288	247
			Surface Daily Totals		1022	1774	1860	1890	1866	1626
Nov. 21	8	4	3.6	54.7	5	17	19	21	22	22
	9	3	11.2	42.7	46	117	129	137	141	135
	10	2	17.1	29.5	83	186	202	212	215	201

	11	1	20.9	15.1	107	227	245	255	258	238
		12	22.2	0.0	115	241	259	270	272	250
			Surface Daily Totals		596	1336	1448	1518	1544	1442
Dec. 21	9	3	8.0	40.9	27	87	98	105	110	109
	10	2	13.6	28.2	63	164	180	192	197	190
	11	1	17.3	14.4	86	207	226	239	244	231
		12	18.6	0.0	94	222	241	254	260	244
			Surface Daily Totals		446	1136	1250	1326	1364	1304

Solar Position and Insolation, 56° N Latitude

	Solar Time		*Solar Position*		*BtuH/Sq. Ft. Total Insolation on Surfaces*					
						South-Facing Surface Angle with Horizon				
Date	*A.M.*	*P.M.*	*Alt.*	*Azm.*	*Horiz.*	*46*	*56*	*66*	*76*	*90*
Jan. 21	9	3	5.0	41.8	11	50	55	59	60	60
	10	2	9.9	28.5	39	135	146	154	156	153
	11	1	12.9	14.5	58	183	197	206	208	201
		12	14.0	0.0	65	198	214	222	225	217
			Surface Daily Totals		282	934	1010	1058	1074	1044
Feb. 21	8	4	7.6	59.4	25	65	69	72	72	69
	9	3	14.2	45.9	65	151	159	162	161	151
	10	2	19.4	31.5	98	215	225	228	224	208
	11	1	22.8	16.1	119	254	265	268	263	243
		12	24.0	0.0	126	268	279	282	276	255
			Surface Daily Totals		740	1640	1716	1742	1716	1598
Mar. 21	7	5	8.3	77.5	28	40	40	39	37	32
	8	4	16.2	64.4	75	119	120	117	111	97
	9	3	23.3	50.3	118	192	193	189	180	154
	10	2	29.0	34.9	151	249	251	246	234	205
	11	1	32.7	17.9	172	285	288	282	268	236
		12	34.0	0.0	179	297	300	294	280	246
			Surface Daily Totals		1268	2066	2084	2040	1938	1700
Apr. 21	5	7	1.4	108.8	0	0	0	0	0	0
	6	6	9.6	96.5	32	14	9	8	7	6
	7	5	18.0	84.1	81	74	66	57	46	29

	8	4	26.1	70.9	129	143	135	123	108	82
	9	3	33.6	56.3	169	208	200	186	167	133
	10	2	39.9	39.7	201	259	251	236	214	174
	11	1	44.1	20.7	220	292	284	268	245	200
	12		45.6	0.0	227	303	295	279	255	209
			Surface Daily Totals		1892	2282	2186	2038	1830	1458
May 21	4	8	1.2	125.5	0	0	0	0	0	0
	5	7	8.5	113.4	25	10	9	8	7	6
	6	6	16.5	101.5	71	28	17	15	13	11
	7	5	24.8	89.3	119	88	74	58	41	16
	8	4	33.1	76.3	163	153	138	119	98	63
	9	3	40.9	61.6	201	212	197	176	151	109
	10	2	47.6	44.2	231	259	244	222	194	146
	11	1	52.3	23.4	249	288	274	251	222	170
	12		54.0	0.0	255	299	284	261	231	178
			Surface Daily Totals		2374	2374	2188	1962	1682	1218
Jun 21	4	8	4.2	127.2	4	2	2	2	2	1
	5	7	11.4	115.3	40	14	13	11	10	8
	6	6	19.3	103.6	86	34	19	17	15	12
	7	5	27.6	91.7	132	92	76	57	38	15
	8	4	35.9	78.8	175	154	137	116	92	55
	9	3	43.8	64.1	212	211	193	170	143	98
	10	2	50.7	46.4	240	255	238	214	184	133
	11	1	55.6	24.9	258	284	267	242	210	156
	12		57.5	0.0	264	294	276	251	219	164
			Surface Daily Totals		2562	2388	2166	1910	1606	1120
Jul. 21	4	8	1.7	125.8	0	0	0	0	0	0
	5	7	9.0	113.7	27	11	10	9	8	6
	6	6	17.0	101.9	72	30	18	16	14	12
	7	5	25.3	89.7	119	88	74	58	41	15
	8	4	33.6	76.7	163	151	136	117	96	61
	9	3	41.4	62.0	201	208	193	173	147	106
	10	2	48.2	44.6	230	254	239	217	189	142
	11	1	52.9	23.7	248	283	268	245	216	165
	12		54.6	0.0	254	293	278	255	225	173
			Surface Daily Totals		2372	2342	2152	1926	1646	1186
Aug. 21	5	7	2.0	109.2	0	0	0	0	0	0
	6	6	10.2	97.0	34	16	11	10	9	7
	7	5	18.5	84.5	82	73	65	68	45	28
	8	4	26.7	71.3	128	140	131	119	104	78

	9		3	34.3	56.7	168	202	193	179	160	126
	10		2	40.5	40.0	199	251	242	227	206	166
	11		1	44.8	20.9	218	282	274	258	235	191
		12		46.3	0.0	225	293	285	269	245	200
			Surface Daily Totals			1884	2218	2118	1966	1760	1392
Sep. 21	7		5	8.3	77.5	25	36	36	34	32	28
	8		4	16.2	64.4	72	111	111	108	102	89
	9		3	23.3	50.3	114	181	182	178	168	147
	10		2	29.0	34.9	146	236	237	232	221	193
	11		1	32.7	17.9	166	271	273	267	254	223
		12		34.0	0.0	173	283	285	279	265	233
			Surface Daily Totals			1220	1950	1962	1918	1820	1594
Oct. 21	8		4	7.1	59.1	20	53	57	59	59	57
	9		3	13.8	45.7	60	138	145	148	147	138
	10		2	19.0	31.3	92	201	210	213	210	195
	11		1	22.3	16.0	112	240	250	253	248	230
		12		23.5	0.0	119	253	263	266	261	241
			Surface Daily Totals			688	1516	1586	1612	1588	1480
Nov. 21	9		3	5.2	41.9	12	49	54	57	59	58
	10		2	10.0	28.5	39	132	143	149	152	148
	11		1	13.1	14.5	58	179	193	201	203	196
		12		14.2	0.0	65	194	209	217	219	211
			Surface Daily Totals			284	914	986	1032	1046	1016
Dec. 21	9		3	1.9	40.5	0	3	4	4	4	4
	10		2	6.6	27.5	19	86	95	101	104	103
	11		1	9.5	13.9	37	141	154	163	167	164
		12		10.6	0.0	43	159	173	182	186	182
			Surface Daily Totals			156	620	678	716	734	722

Source: Morrison, C. A., and Farber, E. A., "Development and Use of Solar Insolation Data in Northern Latitudes for South Facing Surfaces," in Symposium on Solar Energy Applications. New York: ASHRAE, 1974. Reprinted with permission.

Note: Ground reflection is not included in the listed values.

GLOSSARY

absorber The dark surface in a collector that absorbs the solar radiation and converts it to heat energy.

absorptance The ratio of solar energy absorbed by a surface to the solar energy striking it.

active solar system A system that uses mechanical power (fans and/or pumps) to move the collected heat.

altitude The angle of the sun above the horizon.

ASHRAE Abbreviation for the American Society of Heating, Air Conditioning and Refrigerating Engineers.

auxiliary heat The extra heat provided by a conventional heating system for periods of cloudiness or intense cold, when a solar heating system cannot provide enough.

azimuth The angle between solar (true) south and the point on the horizon directly below the sun.

British thermal unit (Btu) The amount of heat needed to raise the temperature of 1 lb. of water 1°F. (Approximately equal to the heat from one wooden kitchen match.)

calorie The amount of heat needed to raise the temperature of 1 gram of water 1°C.

collector A system or device used to collect solar energy and convert it to heat.

collector efficiency The ratio of heat energy provided by a collector to the solar energy striking the collector, expressed in percent.

concentrating collector A device which uses reflective surfaces to concentrate the sun's rays onto a smaller area, where they are absorbed and converted to heat energy.

conduction The transfer of heat energy through a material or from one material to another in contact with it, by the motion of adjacent atoms and molecules.

convection The transfer of heat energy by the movement of liquids or gases (usually air).

degree-day (D.D.) A unit that represents a 1°F difference from 65°F in the mean daily outdoor temperature. The total number of degree-days for a heating season indicates the relative severity of the winter.

delta T A difference in temperature.

design temperature The approximate lowest temperature that can be expected for a certain location. Used to determine the maximum heating demand.

direct gain Direct collection of solar energy, for example, with windows.

double glazing Two layers of glazing with an insulating air space between them.

emittance A measure of the property of a material to emit thermal radiation. Emittance values range from .05 for certain metals to .97 for concrete. Most nonmetals have high values of emittance.

energy The capacity for doing work; including thermal, mechanical, electrical, and chemical.

equinox Either of the two times during the year when the sun crosses the equator and the length of day and night are approximately equal, usually March 21 and September 21.

eutectic salts A type of salts that melt and freeze at approximately room temperatures. Used to store heat.

glazing Translucent or transparent glass or plastic that allows light to pass through.

heat capacity The number of Btu needed to raise the temperature of 1 cubic foot of a material 1°F.

heat exchanger A device, such as a coiled copper tube immersed in a tank of water, that is used to transfer heat from one fluid to another.

heat pump A mechanical device that transfers heat from one medium to another, cooling the first medium and heating the second.

heat storage (thermal storage) Potential heat storage capacity. As the temperature of a material rises, it absorbs (stores) heat, which is given off as the material cools.

infiltration The movement of outdoor air into a structure through cracks around glazing or doors, or through walls, roof, and floor.

infrared radiation Electromagnetic radiation from the sun or a warm body, that has wavelengths longer than visible light.

insolation The total amount of sunlight (direct, diffuse, and reflected) that strikes a surface.

insulation A material that significantly reduces the transfer of heat. Measured in R-value.

langley A measure of solar radiation equal to one calorie per square centimeter.

passive solar system Collection and movement of heat using natural principles, without fans or pumps.

photovoltaics Semiconductor devices that convert sunlight directly into electricity.

radiation The flow of heat by electromagnetic waves, such as light. Heat from the sun comes to earth by radiation.

retrofit To add solar heating or cooling to an existing building.

R-value A unit of thermal resistance. The reciprocal of U-value.

selective surface A material that absorbs most of the sunlight striking it, but emits very little thermal radiation.

specific heat The number of Btu needed to raise the temperature of 1 pound of a material 1°F.

sun path diagram A circular projection of the sky dome onto a flat surface.

temperature stratification The natural formation of thermal layers in a substance where the top layer is warmer than the bottom.

thermal radiation Electromagnetic radiation emitted by a warm body.

thermosiphoning The natural movement that occurs as warm fluid (or gas) rises and cool fluid falls.

ultraviolet radiation Electromagnetic radiation, usually from the sun, with wavelengths shorter than visible light.

U-value (coefficient of heat transfer) The rate of heat loss in Btu per hour through a square foot of surface when the temperature difference between the two sides of the surface is 1°F.

BIBLIOGRAPHY

Anderson, Bruce, and Wells, Malcolm. *Passive Solar Energy*. Andover, Mass.: Brick House Publishing Co., 1981.

Anderson, Bruce, with Riordan, Michael. *The Solar Home Book*. Andover, Mass.: Brick House Publishing Co., 1976.

Anderson, Bruce. *Solar Fundamentals in Building Design*. New York: McGraw-Hill, 1977.

Antolini, Holly, ed. *Solar Heating and Cooling*. Menlo Park, Calif.: Sunset Books, 1978.

ASHRAE, *Handbook of Fundamentals*. New York: American Society of Heating, Refrigerating, and Air Conditioning Engineers, 1967, 1972, and 1977.

Baer, Steve. *Sunspots*. Albuquerque, N.M.: Zomeworks Corporation, 1975.

Bainbridge, David. *The Integral Passive Solar Water Heater Book*. Berkeley, Calif.: The Passive Solar Institute, 1981.

Bainbridge, David. *The 2nd Passive Solar Catalogue*. Berkeley, Calif.: The Passive Solar Institute, 1980.

Bainbridge, David; Corbett, J.; and Hofacre, J. *Village Homes' Solar House Designs*. Emmaus, Penn.: Rodale Press, 1979.

Bryenton, Roger; Cooper, Ken; and Mattock, Chris. *The Solar Water Heater Book*. Toronto, Ont.: Renewable Energy in Canada, 1980.

Butti, Ken, and Perlin, John. *A Golden Thread: 2,500 Years of Solar Architecture and Technology*. Palo Alto, Calif.: Cheshire Books, 1980.

Carter, Joe, ed. *Solarizing Your Present Home*. Emmaus, Penn.: Rodale Press, 1981.

Clegg, Peter, and Watkins, Derry. *The Complete Greenhouse Book*. Charlotte, Vt.: Garden Way Publishing, 1978.

Davis, N. D., and Lindsey, L. *At Home in the Sun: An Open House Tour of Solar Homes in the United States*. Charlotte, Vt.: Garden Way Publishing, 1979.

Eccli, Eugene. *Low Cost Energy Efficient Shelter for the Owner and Builder*. Emmaus, Penn.: Rodale Press, 1975.

Fisher, Rick, and Yanda, William. *The Food and Heat Producing Solar Greenhouse*. Santa Fe, N.M.: John Muir Publications, 1976.

Langdon, William. *Movable Insulation*. Emmaus, Penn.: Rodale Press, 1980.

Mazria, Edward. *The Passive Solar Energy Book*. Emmaus, Penn.: Rodale Press, 1979.

McCullagh, James C. *The Solar Greenhouse Book*. Emmaus, Penn.: Rodale Press, 1978.

Olgyay, Victor. *Design with Climate*. Princeton, N.J.: Princeton University Press, 1963.

Reif, Daniel K. *Solar Retrofit: Adding Solar To Your Home*. Andover, Mass.: Brick House Publishing Co., 1981.

Shurcliff, William A. *New Inventions in Low-Cost Solar Heating.* Andover, Mass.: Brick House Publishing Co., 1979.

Shurcliff, William A. *Superinsulated Houses and Double-Envelope Houses.* Andover, Mass.: Brick House Publishing Co., 1981.

Shurcliff, William A. *Thermal Shutters and Shades.* Andover, Mass.: Brick House Publishing Co., 1980.

Sussman, Art, and Frazier, Richard. *Handmade Hot Water Systems.* Point Arena, Calif.: Garcia River Press, 1978.

Wagner, W. *Modern Carpentry.* South Holland, Ill.: Goodheart-Willcox Company, 1976.

Wells, Malcolm and Spetgang, I. *How To Buy Solar Heating . . . Without Getting Burnt.* Emmaus, Penn.: Rodale Press, 1978.

Magazines of interest:

Solar Age, Church Hill, Harrisville, N.H.

New Shelter, 33 E. Minor Street, Emmaus, Penn.

INDEX